T0192404

Linux

Linux is one of the most widely used operating systems. It was created to provide a free or low-cost operating system for personal computer users. Linus Torvalds published Linux on September 17, 1991, and it was written in the C programming language. It has since earned a reputation for being a high-performing and efficient system. This is a fairly comprehensive operating system that includes a graphical user interface (GUI), TCP/IP, the Emacs editor, and the X Window System, among other features.

Debian, Ubuntu, Fedora, Red Hat Linux, SUSE Linux, Gentoo, Kali Linux, and Linux Mint are some of the finest Linux distributions. Linux is a very popular operating system today because of features such as multiuser operating system management, multitasking paradigm, multiprogramming concepts, and virtual memory. Many corporations and individuals, as well as firms such as Canonical, use Linux for their servers because of security concerns and positive feedback from the user community. Linux is also used in mobile devices, smart TVs, and so on.

Key Features:

- A step-by-step approach to problem solving and skill development

- A quick run-through of the basic concepts, in the form of a "crash course"

- An advanced, hands-on core concepts, with a focus on real-world problems

- An industry-level coding paradigm, practice-oriented explanatory approach

- A special emphasis on writing clean and optimized code, with additional chapters focused on coding methodology

Linux

The Ultimate Guide

Sufyan bin Uzayr

CRC Press
Taylor & Francis Group
Boca Raton London New York

CRC Press is an imprint of the
Taylor & Francis Group, an **informa** business

First edition published 2023
by CRC Press
6000 Broken Sound Parkway NW, Suite 300, Boca Raton, FL 33487-2742

and by CRC Press
4 Park Square, Milton Park, Abingdon, Oxon, OX14 4RN

CRC Press is an imprint of Taylor & Francis Group, LLC

Library of Congress Cataloging-in-Publication Data

Names: Bin Uzayr, Sufyan, author.
Title: Linux : the ultimate guide / Sufyan bin Uzayr.
Description: First edition. | Boca Raton : CRC Press, 2023. | Includes
bibliographical references and index.
Identifiers: LCCN 2022025683 (print) | LCCN 2022025684 (ebook) | ISBN
9781032312255 (hardback) | ISBN 9781032312248 (paperback) | ISBN
9781003308676 (ebook)
Subjects: LCSH: Linux. | Operating systems (Computers)
Classification: LCC QA76.774.L46 B56 2023 (print) | LCC QA76.774.L46
(ebook) | DDC 005.4/46--dc23/eng/20220920
LC record available at https://lccn.loc.gov/2022025683
LC ebook record available at https://lccn.loc.gov/2022025684

ISBN: 9781032312255 (hbk)
ISBN: 9781032312248 (pbk)
ISBN: 9781003308676 (ebk)

DOI: 10.1201/9781003308676

Typeset in Minion
by Deanta Global Publishing Services, Chennai, India

Contents

Acknowledgments

There are many people who deserve to be on this page, for this book would not have come into existence without their support. That said, some names deserve a special mention, and I am genuinely grateful to:

- My parents, for everything they have done for me.

- My siblings, for helping with things back home.

- The Parakozm team, especially Divya Sachdeva, Jaskiran Kaur, and Vartika, for offering great amount of help and assistance during the writing of this book.

- The CRC team, especially Sean Connelly and Danielle Zarfati, for ensuring that the book's content, layout, formatting, and everything else remains perfect throughout.

- Reviewers of this book, for going through the manuscript and providing their insight and feedback.

- Typesetters, cover designers, printers, and everyone else, for their part in the development of this book.

- All the folks associated with Zeba Academy, either directly or indirectly, for their help and support.

- The programming community in general, and the web development community in particular, for all their hard work and efforts.

Sufyan bin Uzayr

Author

SUFYAN BIN UZAYR IS a writer, coder, and entrepreneur with more than a decade of experience in the industry. He has authored several books in the past, pertaining to a diverse range of topics, ranging from history to Computers/IT.

Sufyan is the Director of Parakozm, a multinational IT company specializing in EdTech solutions. He also runs Zeba Academy, an online learning and teaching vertical with a focus on STEM fields.

Sufyan specializes in a wide variety of technologies, such as JavaScript, Dart, WordPress, Drupal, Linux, and Python. He holds multiple degrees, including ones in management, IT, literature, and political science.

Sufyan is a digital nomad, dividing his time between four countries. He has lived and taught in universities and educational institutions around the globe. Sufyan takes a keen interest in technology, politics, literature, history, and sports, and in his spare time he enjoys teaching coding and English to young students.

Learn more at sufyanism.com.

CHAPTER 1

Desktop Environments for Linux

IN THIS CHAPTER

➢ Desktop environments for linux

➢ History of desktop environment

This chapter will cover the fundamentals of the desktop environment for Linux with its significant concepts, primary usage, and more. So let's begin with the introduction of the desktop environment (DE). But in the coming chapter, we will discuss some of the valuable desktop environments of Linux in detail.

DESKTOP ENVIRONMENT INTRODUCTION

A desktop environment implements the desktop metaphor of a bundle of programs running on top of an operating system that shares a standard graphical user interface (GUI). Sometimes, it is described as a graphical shell. The desktop environment mainly was on personal computers until mobile computing. Desktop GUIs help the user quickly access and edit files, while they usually don't provide access to all of the features in the underlying operating system. Besides, the traditional command-line interface (CLI) is still used when complete control over the operating system is required.

DOI: 10.1201/9781003308676-1

It typically consists of various icons, windows, toolbars, folders, wallpapers, and desktop widgets. A GUI also provides drag and drop functionality and other features that complete the desktop metaphor. A desktop environment aims to be a way for the user to interact with the system using concepts similar to those used to interact with the rest of the world, such as buttons and windows.

While the term desktop environment is described initially as a style of user interface given by the desktop metaphor, it has also defined the programs that realize the metaphor. The usage has been popularized by projects such as KDE Plasma, GNOME, XFCE, MATE, Budgie, Cinnamon, and LXDE.

LINUX

Like other operating systems such as Windows, iOS and Mac OS, Linux is an operating system. One of the world's most popular platforms, Android is powered by a Linux operating system. An application is a software that controls all hardware resources associated with your desktop or laptop. To put it simply, the operating system controls the connection between your software and your hardware. Without an operating system (OS), the software will not work.

Components of Linux Application

- **Bootloader**: The software that controls the process of launching your computer. It will simply be a splash screen that pops up and eventually moves to the operating system for many users.

- **Kernel**: The kernel is the system's core and controls CPU, memory, and border-related devices. The kernel is a very low OS rate.

- **Init System**: This sub-system initiates user space and is charged by control daemons. It is an init program that controls the startup process, when the initial boot has been transferred to the bootloader.

- **Daemons**: These are background services (printing, sound, editing, etc.) that may start during launch or after the desktop entry.

- **Graphics Server**: This is a sub-system that displays graphics on your monitor. It is usually called an X server or just an X.

- **Desktop Environment**: This is the piece users are interacting with. There are many desktop areas (GNOME, Cinnamon, Mate,

Pantheon, Enlightenment, KDE, Xfce, etc.). Each desktop includes built-in applications (such as file managers, configuration tools, web browsers, and games).

- **Applications**: Desktop locations do not provide a complete network of applications. Like Windows and macOS, Linux offers thousands of software titles that are easily accessible and installed. Many modern Linux distributions include tools like the App Store that integrates and simplifies system installation.

THE DESKTOP ENVIRONMENT IN LINUX

The first "desktops" on Linux were not yet desktops. Instead, they were window managers using the X Window System. X provided basic building blocks with visual effects, such as making windows on the screen and giving keyboard and mouse input. To make the graphical X space usable, you need a way to manage all the windows in session. Using the X program as xterm or xclock opens that program in a window. The window manager traces the windows and performs essential house maintenance, allowing you to move the windows and minimize them. The rest is up to you. You could start programs when X starts by listing them in the ~ / .xinitrc file, but in most cases, you could run new programs from xterm.

There are the following terms used: graphical user interface, command-line interface icons, windows, toolbars, folders, wallpapers and desktop widgets, elements of graphical user interfaces (GUI), and WIMP. Let's discuss the following terms.

Graphical User Interface

It is an interface that allows interaction with devices through graphical icons and an audio indicator such as notation instead of text-based user interfaces, typed commands, and text navigation. In reaction to command-line interfaces' sensed steep learning curve, GUIs require typing commands on a computer keyboard.

The actions in a GUI are performed via direct manipulation of the graphical elements. GUIs are used in mobile devices such as audio MP3 players, portable players, gaming devices, smartphones, household, office, and industrial controls. The GUI is not to be applied to the lower-display resolution interfaces, such as video games; this term is restricted to the scope of two-dimensional display screens to describe generic information of the scientific research at the Xerox Palo Alto Research Center.

The temporal behavior of a GUI and designing the visual composition is an essential part of software application programming in human–computer interaction. Its goal is to enhance ease of use for the underlying logical design of a stored program, a design discipline named usability. User-centered design methods are used to ensure that the visual language introduced in the design is well-tailored to the tasks.

The visible graphical interface features are sometimes referred to as chrome or GUI. The users interact with information by visual widgets that allow interactions to the kind of data they hold. The widgets having a well-designed interface are selected to support these necessary actions to achieve users' goals. A model view controller allows flexible structures in which the interface is independent of and indirectly linked to application functions so that the GUI can be customized easily. It will enable the user to select a different skin and eases the designer's work to change the interface as the user needs to evolve. Good interface design relates to users more and system architecture less. Large widgets, such as windows, provide a frame or container for the main presentation content, such as a web page email message.

A GUI is designed for the requirements of a market as application-specific graphical user interfaces. Examples are automated teller machines (ATM), point-of-sale (POS) touch screens at restaurants, self-service checkouts used in a retail store, airline self-ticket and check-in, a train station or a museum, and monitors or control screens in an embedded industrial application which employ a real-time operating system (RTOS).

Examples

The following are examples of the graphical environments for Linux.

- Ambient

- Bugie Desktop

- Budgie

- CDE

- Cinnamon

- Cutefish

- Deepin DE

- EDE

- Elokab
- Enlightenment
- Étoilé
- GNOME Shell
- GNUstep
- Innova
- Katana
- KDE Plasma 5
- Liri Shell
- Lumina
- LXDE
- LXQt
- MATE
- MaXX
- Maynard
- Mezzo
- Moksha
- Pantheon
- Project Looking Glass
- oZone GUI
- Razor-qt
- ROX Desktop
- Sugar
- theShell
- Trinity
- UKUI (desktop environment)

- UKUI (desktop environment)

- Unity

- vera

- Xfce

Parts of the Graphical User Interface

The GUI uses a combination of technology and devices to provide a platform that users can interact with to collect and generate information. A series of elements associated with visual language has been developed to represent information stored on computers. It makes it easier for people with few computer skills to work and use computer software. The most common combination of such things in GUIs is a window, icons, menus, pointer (WIMP) paradigm, especially on personal computers.

The WIMP style of interaction uses a visual input device to represent the location of the device's visual interface, usually a mouse, and presents window layouts and is represented by icons. Available commands are merged in menus, and actions are performed by tapping. A window manager facilitates the interaction between windows, applications, and the installation window. The Windows installation system manages hardware devices such as pointing devices, image hardware, and cursor position.

On computers, all of these features are modeled using a desktop metaphor to produce a simile called a desktop location where the display represents a desktop, where text and folders can be placed. Window managers and other software come together to mimic a desktop environment with varying levels of virtual reality.

Entries may appear in the list to make text and details or in the integration grid with large icons with minimal space under the text. Variations exist, such as the multi-column list and the object grid with text lines extending sideways from the thumbnail. Multi-row buildings and multiple columns are commonly found on the web "shelf" and "waterfall." The former is found in image search engines, where images appear with fixed but variable lengths and are often used in CSS format and parameter display: inline-block. The waterfall structure found in Imgur and Tweetdeck with a fixed width, but variable length for each item is usually used to specify column width.

Post-WIMP Interface

Small app mobile devices such as digital assistants (PDAs) and smart-phones often use WIMP features with various metaphors that combine due to the space limitations and input devices available. WIMP-incompetent applications can use new interactive methods, collectively called the post-WIMP user interface.

Since 2011, some touch screen operating systems such as Apple's iOS (iPhone) and Android use a GUI class named post-WIMP. These interactive support systems use more than one finger connected to the display, allowing actions such as compression and rotation, which a single mouse and mouse do not support.

List of Graphical User Interface

Graphics elements are those elements used by GUIs to provide a consistent visual language representing the information stored on computers. It makes it easier for people with few computer skills to operate and use computer software.

This section describes the most common features of visual language links found in the WIMP paradigm, which stands for "window, icon, menu, cursor," although many are used in another graphical post-WIMP interface. These features are usually integrated using a widget toolkit or desktop area.

STRUCTURAL ELEMENTS OF DE

The visual user interface uses visual principles to represent the general information displayed. Some principles are used to create a strand of objects in which the user can interact and define the interface's appearance.

Window

A window is a screen area that displays information, the content of which is displayed independently across the screen. An example of a window appears on the screen when "My Documents" or any other icon is clicked on the Windows Operating System. It is easy for the user to trick the window: it can be displayed and hidden by clicking on the icon or app, and it can be moved to any location by dragging it (i.e., by clicking on a window area – usually the title bar at the top – and keeping the pointing device button pressed, then moving the pointing device). A window can be placed in front of or behind another window, its size can be adjusted, and scroll bars can be used to navigate through sections within it. Many windows can reopen at the same time, where each window can display a separate application or file. This is very useful when working as a multi-tasker. System memory is the limit of the number of windows that can be opened simultaneously. There are also many types of special windows.

- The container window closes some windows or controls. When moving or resizing, locks move, resize, rearrange, or cut the container window.

- The browser window allows the user to view and navigate a collection of items, like files or web pages. Web browsers are an example of the types of windows.

- The text terminal windows introduce character-based text interaction, commanded between the entire image area. MS-DOS and UNIX consoles are examples of the types of windows. Terminal windows are often compliant with hotkey and CRT-based terminal display systems that precede GUIs, such as the VT-100.

- The child window opens automatically or due to user activity in the parent window. Windows that pop up online can be child windows.

- A message or dialog box is a child window type. These are usually small and basic windows opened by the user information display system and the user information. They probably always have one or more buttons, which allow the user to tick the box with positive, negative, or intermediate feedback.

Thumbnails (icon)

Thumbnail (icon) is a small image representing objects such as a file, program, web page, or command. They are a quick way to issue commands, open documents, and run programs. Thumbnails are also very useful when searching for an item in a browser list because, in most applications, all documents using the same extension will have the same icon.

Controls (or Widgets)

Visual interface components such as image control elements, controls, or widgets are part of the software that a computer user uses trickery to read or edit information about the application. Each widget facilitates some user–computer interaction. Editing user interaction with the widget tool kit allows developers to re-use the same function code. It gives users a common language to work with, maintaining consistency throughout the information system.

Common uses of widgets include the display of related clusters (such as various lists and canvas controls), implementation of actions and processes within the interface (buttons and menus), roaming within the information system area (links, tabs, and scroll bars), and representing and decrypting data values (e.g., radio buttons, sliders, and spinners).

Tabs

A tab is usually a small rectangular box containing a text label or an image icon associated with a viewing window. When the view window is activated, it shows the widgets with that tab. It can also group tabs that allow users to switch between different widgets quickly. It applies to all modern web browsers. With these browsers, you could have multiple web pages open at once in a single window and quickly navigate through them by clicking the tabs associated with the pages. Tabs are usually grouped at the top of the window but may also be grouped to the side or bottom of the window. Tabs are also present in the settings of many application settings. Microsoft Windows, for example, uses the tabs in most of its control panel discussions.

Menu

It allows the user to execute commands just by selecting from the options menu. Options are set with the mouse or other device pointing within the GUI. The keyboard can also be used. The menus are appropriate because they indicate which commands are available within the software. It reduces the number of documents a user reads to understand the software.

- The menu bar is displayed horizontally at the top of the screen and overall windows. The drop-down menu is usually associated with this type of menu. When you click on a menu option, a drop-down menu will appear.

- The menu has a visual title within the menu bar. Its content is displayed only when the user selects it by the cursor. The user is then able to select items within the drop-down menu. When a user clicks elsewhere, the menu content will disappear.

- The context menu is not visible until the user performs a specific mouse action, such as pressing the right mouse button. A menu will appear under the cursor when the software-specific mouse action occurs.

- Menu extensions are separate items within or next to the menu.

OTHER COLLABORATIVE ELEMENTS

Some common interaction expressions appear in the visual language used in the GUIs. Interactive elements are interactive objects that represent a state of continuous activity or modification, such as visual reminders of a user's interface.

Cursor

It disables the indicator used to display location on a monitor or other display device that responds to input from a text input or target device.

Identifier

The cursor echoes the movement of the pointing device, usually a mouse or touchpad. The cursor is where the actions start with direct touches, such as clicking, tapping, and dragging.

Input Point

A caret, text cursor, or input point represents the point of use seen where the focus is. It represents an item used as the default theme for user-initiated commands such as text typing, selected startup, or copy-paste function.

Choice

The selection is a list of things in which the user's performance will occur. The user usually adds items to the list in person, although the computer may create a selection automatically.

Repair Handle

The handle indicates the starting position of the drag and drops operation. Usually, the cursor's position changes when placed on the handle, indicating an icon representing the supported drag function.

HISTORY OF DESKTOP ENVIRONMENT

The first desktop space was created by Xerox and sold with Xerox Alto in the 1970s. Xerox generally regarded alto as a personal office computer; it failed on the market due to poor marketing and high value.

The desktop analogy was introduced to commercial computers by the original Macintosh from Apple in 1984 and has been popular with Windows from Microsoft since the 1990s. As of 2014, the most popular desktop areas are a descendant of these previous sites, including the Windows shell used on Microsoft Windows and the Aqua environment used for macOS. Compared to X-based desktop environments found in applications such as Unix such as Linux and FreeBSD, Windows and macOS desktop environments have limited layouts and static features, with integrated "seamless" designs intended to provide especially the consistent customer sensitivity throughout the installation.

Microsoft Windows dominates the market share between personal computers and desktops. Computers using Unix operating systems such as macOS, Chrome OS, Linux, BSD, or Solaris are less common. Since 2015, there has been a growing market for less expensive Linux PCs using the X Window System or Wayland. With a wide selection of desktop locations. Among the most popular are Chromebooks and Google Chromeboxes, Intel's NUC, Raspberry Pi, etc.

The situation is the same for tablets and smartphones, as there are apps like Unix that dominate the market, including iOS (BSD based), Android, Tizen, Sailfish, and Ubuntu (all available on Linux). Microsoft Windows Phone, Windows RT, and Windows 10 are used on a minimal number of tablets and smartphones. However, most applications such as Unix hosted on portable devices do not use the X11 desktop areas used by other operating systems such as Unix instead of relying on communications based on different technologies.

Desktop Environment for the X Window System

For applications that use the X Window System (usually Unix family systems such as Linux, BSD, and the official UNIX distribution), desktop environments are highly flexible and customized to meet users' needs. In this context, the desktop area usually contains a few different components, including a window manager (such as Mutter or Kwin), a file manager (such as Files or Dolphin), a set of graphic themes, and tools (like GTK + and Qt) and desktop management libraries. These individual modules can be customized and configured to suit users, but most desktop environments offer automatic configuration that works with minimal user settings. So now look at what exactly X Window System is, as given below.

X Window System

The X Window, also known as X11, or X, installs bitmap display windows, standard operating systems such as Unix. It is Unix-family systems such as Linux, the BSDs, and formal UNIX distributions. The X provides the basic framework of the GUI: drawing and moving windows on a display and interactive mouse and keyboard devices. X does not authorize user interaction – individual programs manage this. Thus, the visual style of X-based locations varies greatly; different systems may present very different connections.

X was introduced as part of Project Athena at the Massachusetts Institute of Technology (MIT) in 1984. The X Protocol has been in version 11 (hence the "X11") since September 1987. The X.Org Foundation is leading the X project, with the current use of the reference, X.Org Server, available as free software and open sources under MIT License and similar licensing licenses.

The X was specially designed for use in network connectivity instead of a virtual or attached display device. X includes network visibility, which means that the X-based computer application somewhere in the network (such as the Internet) can display its user interaction on an X server running on another computer in the network. The X server is usually a provider of graphics and keyboard/mouse events for X clients, meaning that the X server usually runs on a computer in front of a human user. In contrast, the X client applications run anywhere on the network and communicate with them. A user's computer can request the provision of image content and receive events on input devices, including keyboards and mice.

X Window System Software Architecture

It uses a client-server model: the X server communicates with various client programs. The server accepts requests for image extraction (windows) and returns user input (from keyboard, mouse, or touch screen).

An application displayed in a window is another display system, a program that controls video output to a PC, a special piece of hardware. This client name – the end-user, the server, and client applications – often confuses new X users because the names seem undone. But X takes the view of the application instead of the end-user: X provides the display and I/O services in applications, so it is a server; applications use these services, so they are customers. The communication protocol between server and client works across the network: the client and server may work on the same or different machine, possibly with other formats and applications. The X client can mimic the X server by providing display services to other clients. It is known as the "X nesting." Open-source clients such as Xnest and Xephyr support such X breeding.

User Interface

X defines most protocols and original images. It does not deliberately contain specifications of the user interface, such as the button, menu, or window title window styles. Instead, application software such as window managers, GUI widget tools and desktop environments, or user-specific graphical user interfaces define and provide such information. As a result, there is no standard X interface, and several desktop areas are already popular among users.

A window manager can control the appearance of application windows. It may result in desktop communications reminiscent of those for Microsoft Windows or Apple Macintosh (examples include GNOME 2, KDE, Xfce) or have very different controls (such as a tile window manager, like wmii or Ratpoison). Some optical connectors such as Sugar or Chrome OS avoid desktop nicknames, making their connections to special applications easier. Window managers vary in complexity and sophistication from bare-bones (e.g., twm, X-window primary window manager, or evilwm, over-the-window window) to more expansive desktop areas such as lighting and even direct use. Windows-direct market manages as a trading platform.

Most users use an X with a desktop area, which installs various applications using a fixed user interface in addition to the window manager. Popular desktop areas include GNOME, KDE Plasma, and Xfce. The UNIX 98 General Area is the Common Desktop Environment (CDE). The Freedesktop.org initiative discusses the interaction between desktops and the necessary components of a competitive X desktop.

HISTORY OF X WINDOW SYSTEM

Several bitmap display systems preceded X. To Xerox came Alto (1973) and Star (1981). From Apollo Computer came the Display Manager (1981). From Apple came Lisa (1983) and Macintosh (1984), Unix World's Andrew Project (1982) and Rob Pike's Blit (1982) terminology. Carnegie Mellon

University has developed a remote access app called Alto Terminal, which displays scattered windows on Xerox Alto. It makes remote visitors (usually DEC VAX systems using Unix) responsible for hosting window display and window content events, refreshing as needed.

X gets its name as a follower of the pre-1983 window system called W (the letter preceding X in English characters). It operates under a V operating system. Using a network protocol that supports terminals and image windows, the server stores the display list.

Competitors

Some people have tried to write alternatives for X. Other methods include Sun's NewS and NeXT's Display PostScript, both PostScript-based systems that support unexplained side view processes, X did not. Other current options include:

- MacOS uses its own windows program. When Apple Inc. bought NeXT to build Mac OS X, it changed the Display PostScript to Quartz. One of Quartz's authors explained that if Apple had added support for all the features it wanted to include in the X11, it would not be very similar to the X11 or compatible.

- Android uses the Linux kernel and uses its system to decrypt an interface called SurfaceFlinger.

- A few X.Org engineers are developing Wayland as an alternative to X. It works directly with GPU hardware. Wayland can use the X.org server as a client, which needs to be root-free.

X Window System

Some window managers in X Window System like IceWM, Fluxbox, Openbox, ROX Desktop, and WindowMaker contain highly degraded desktop features, such as an integrated local file manager. In contrast, others, such as evilwm and wmii, do not provide such features. Not all program codes are part of the desktop area with visible effects. One of you may be a low-level code. KDE, for example, provides so-called KIO slaves that give the user access to a wide variety of visual devices.

Examples of Desktop Environments

The most common desktop space for personal computers is Windows Shell on Microsoft Windows. Microsoft made significant efforts to make the Windows shell look fun. As a result, Microsoft introduced theme support on Windows 98, various Windows XP visual styles, Aero version in Windows Vista, Microsoft design language (coded "Metro") on Windows 8, and Fluent Design System and Windows Spotlight on Windows 8. The Windows shell can be expanded using Shell extensions.

Typical desktop environments for apps like Unix use the X Window System. They include KDE, GNOME, Xfce, LXDE, and Aqua, any of which can be user-selected and not limited to the operating system. Many other desktop areas are also available, including (but not limited to) CDE, EDE, GEM, RIXX Interactive Desktop, Sun's Java Desktop System, Jesktop, Mezzo, Project Looking Glass, ROX Desktop, UDE, Xito, and Xfast. In addition, there is the FVWM-Crystal, which contains a powerful configuration of the FVWM window manager theme and adds totally to create a "build kit" to create a desktop space.

X's window controls that are intended to be used independently – except elsewhere in the desktop – also include features found in common desktop areas, which are very bright. Other examples include OpenBox, Fluxbox, WindowLab, and Fvwm, as well as WindowMaker and AfterStep, both combining the look of the NEXTSTEP GUI. However, newer versions of other operating systems make it ready.

VARIOUS DESKTOP ENVIRONMENTS

Ambient

Ambient is an MUI-based desktop space for MorphOS. Its development was started in 2001 by David Gerber. Its main objectives were to be fully asynchronous, fast, and straightforward. The Ambient remotely resembles Workbench and Directory Opus Magellan, which seek to integrate the best of both worlds.

Features

- Arexx writing language support

- A fully compatible, multi-threaded design

- Instant not synced I/O instant file functions and file notifications

- Support for PNG and other Amiga icon formats

- Built-in icon, workspace, and wbstart libraries

- Built-in applications such as disk formatting and asset manager

- Panels used as program launchers

Bugie Desktop

Budgie is a desktop site currently using GNOME technology similar to GTK (> 3.x), developed by the Solus project and contributed by multiple communities such as Arch Linux, Manjaro, openSUSE Tumbleweed, and Ubuntu Budgie. Budgie's design emphasizes simplicity, minimalism, and elegance. The Solus Project will replace the GTK library with the Enlightenment Foundation Library (EFL) to release Budgie 11. Budgie was initially developed as a default desktop distribution platform for Evolve OS Linux.

CDE

Common Desktop Environment (CDE) is a Unix desktop and OpenVMS desktop, based on the Motif widget toolkit. It was part of the UNIX 98 Workstation Product Standard and has long been a Unix desktop associated with Unix commercial workplaces. After a long history as proprietary software, CDE was released as free software on August 6, 2012, under the GNU Lesser General Public License, version 2.0 or later. Since its release as free software, CDE has been installed on Linux and BSD alternatives.

Cinnamon

Cinnamon is a free and open-source desktop X Window System sourced from GNOME 3, following standard desktop metaphor agreements. Cinnamon is the main desktop distribution platform for Linux Mint and is available as a desktop of your choice for other Linux distributions and other applications such as Unix.

The development of Cinnamon began in the April 2011 release of GNOME 3 when the standard desktop GNOME 2 desktop was left in favor of GNOME Shell. Following several attempts to extend GNOME 3 to suit the design goals of Linux Mint, Mint developers have installed several GNOME 3 components to create a standalone desktop space. The split on GNOME was completed on Cinnamon 2.0, released in October 2013. Apples and desktops are no longer compatible with GNOME 3.

As a distinguishing feature of Linux Mint, Cinnamon has generally received good media coverage, mainly due to its ease of use and soft learning curve. In terms of its sequential design model, Cinnamon is similar to the Xfce desktop and GNOME 2.

Enlightenment

Light, also known as E, is a compact window manager for the X Window System. From version 20, Enlightenment is also the creator of Wayland. Light developers have dubbed it the "real eye candy window manager."

Enlightenment includes image shell rendering functions and can be used with programs written for GNOME or KDE. Used in conjunction with the Enlightenment Foundation Libraries (EFL), Lighting can refer to the entire desktop area.

MATE

MATE is a free and open-source desktop software that works on Linux, BSD, and illumos applications. The name initially was all uppercase letters following the nomenclature of other Free Software desktop sites like KDE and LXDE. The repetitive backronim "MATE Advanced Traditional Environment" was adopted by the majority of the MATE community. Using a new name avoids conflict of words with parts of GNOME 3.

KDE Plasma 5

KDE Plasma 5 is the current generation of graphics software created by KDE, especially for Linux applications. KDE Plasma 5 followed KDE Plasma 4 and was released on July 15, 2014. It includes a new automatic theme, known as "Breeze," and increased integration across all different devices. Image integration is fully integrated into QML, which uses OpenGL to accelerate hardware, resulting in better performance and reduced power consumption. Mobile is a version of the Plasma 5 Linux-based smartphones.

LXDE

LXDE is a free desktop space with relatively low service requirements. It is particularly suitable for use on older desktop computers such as netbooks or systems-on-chip computers. LXDE is written in C programming language, using the GTK 2 tool kit, and works on Unix and other POSIX compliant platforms, such as Linux and BSD. The LXDE project aims to provide faster and more powerful desktop space. In 2010, experiments suggested that LXDE 0.5 had the lowest memory usage of the four most popular desktop devices (GNOME 2.29, KDE Plasma Desktop 4.4, and Xfce 4.6). It consumed less power, suggesting that Linux-distributed mobile computers use LXDE 0.5 to discharge their batteries at a slower speed than those at other desktop locations.

In the chapters that follow, you will get a deeper knowledge of the DE's of Linux.

CHAPTER SUMMARY

Desktop environments are an integral part of the Linux desktop, while Linux servers often rely on a command-line interface. It is not that you cannot install desktop environments on Linux servers. Still, it wastes valuable system resources that can be used by applications running on the server. You will have a little better understanding of desktop environments on Linux now. We recommend reading the explanation of Linux and why there is so much Linux distribution.

CHAPTER 2

KDE Plasma Desktop Environment

IN THIS CHAPTER

- ➤ Introduction
- ➤ KDE history
- ➤ KDE applications
- ➤ KDE version history
- ➤ Installation
- ➤ Features

After a brief introduction of desktop environment in the previous chapter, we will start in this chapter with our first desktop Linux-based operating systems named KDE. Primarily, it is an official Linux OS and has various KDE features. KDE is based on the pure KDE built from the Ubuntu repositories. The first release was Beta 1 on October 20, 1997. Three additional Betas followed on November 23, 1997, February 1, 1998, and April 19, 1998.

INTRODUCTION

There are various terms to discuss to understand the concept of the Ubuntu KDE. So let's begin this with Ubuntu, and then we shall move

DOI: 10.1201/9781003308676-2

forward to the desktop environment KDE. Now we are going to cover basic terms before going deep into the KDE desktop environment such as distribution, open-source Linux desktop environment, GUI, TUI, CLI, and so on.

What Is Distribution?

The term "distribution" refers to the combination of these packaging of the kernel with the GNU libraries and applications. Ubuntu is one such distribution. It contains the Linux kernel, the GNU tools, and many other applications and libraries.

Open-Source Linux Desktop Environment

The word "Open-Source" is attributed to the Linux community which brought it into existence along with the introduction of Linux. "Linux" came into existence only based on kernel. Many people and communities started contributing toward making it a complete operating system which could replace UNIX.

Free Software

"Free software" is software that respects users' freedom and community. Approximately, it means that the users have the freedom to do anything such as run, copy, distribute, study, change, and improve the software. Therefore, "free software" is a topic of liberty, not price. A program is a free software that adequately gives users all of these freedoms. Otherwise, it is not free.

Key Points

- The freedom to run the program as per your wish
- Free software can be commercial
- The freedom to get the source code and make changes
- Legal considerations
- Contract-based licenses

Next, we will discuss the terms GUI, CLI, and TUI, which are also related to the Ubuntu desktop environment KDE.

This section examines the GUI and the significant components of the Linux GUI. You will learn about standard window managers and desktop environments used with Linux.

GRAPHICAL USER INTERFACE

GNOME is the default GUI for most Ubuntu installations and is (loosely) based on the Apple ecosystem. A GUI or graphical application is anything you can interact with using your mouse, touchpad, or touch screen. You have various icons and other visual prompts that you can activate with a mouse pointer to access the functionalities. DE provides the graphical user interface to interact with your system. You can use GUI applications such as GIMP, VLC, Firefox, LibreOffice, and file manager for various tasks.

Features of Linux GUI

The interface allows users to interact with the system visually with icons, windows, or graphics in a GUI. The kernel is the heart of Linux, whereas GUI is the face of the operating system provided by the X Window System or X.

The product of the X.Org Foundation, an open-source organization, X Window System, is a protocol that allows links to be built on their X Server. You can use the X in one of the many window managers or desktop environments, such as the GNU Network Object Model Environment (GNOME) or the Kool Desktop Environment (KDE). The desktop space includes a window manager and is a much more integrated system than a window manager. Built on a window manager requires both X Windows and a window manager.

Features of a GUI

There are unique features and tools to interact with the software to make the GUI easy to use. Below is a list of all of these with a brief description.

- **Button:** A graphical representation of a button that acts as a program when pressed.

- **Dialog Box:** The window type displays additional information and asks the user for input.

- **Thumbnail:** It is a small representation of a program image, feature, or file.

- **Menu:** A list of commands or options provided by the user through the menu bar.

- **Menu Bar:** It is a small, horizontal bar containing menu labels.

- **Ribbon:** Set up a file menu and toolbar that integrates program functions.

- **Tab:** A clickable area at the top of a window shows another page or location.

- **Toolbar:** The Button Bar, usually near the app window's top, controls software operations.

- **Window:** A rectangular section of a computer display that shows the operating system.

The GUI uses icons, windows, and menus to execute commands, such as opening, deleting, and moving files. Although the GUI operating system is navigated using the mouse, the keyboard can also use with keyboard shortcuts or arrow keys.

For example, if you wanted to open an application on the GUI system, you could move the mouse pointer to the system icon and double-click it. With the command-line interface, you will need to know the commands to go to the program's directory, enter the list of files, and then use the file.

Benefits of GUI

A GUI is considered more user-friendly than a text-based command-line interface, such as MS-DOS, or the shell of operating systems like UNIX. Unlike command-line or CUI operating systems, such as UNIX or MS-DOS, GUI operating systems are easy to read and use because commands do not need to be memorized. Additionally, users do not need to know any programming languages. Thanks to its ease of use and modern appearance, GUI operating systems dominate today's market.

Command-Line Interface

CLI is a command-line program that accepts inputs to perform a particular function. Any application you can use via commands in the terminal falls into this category. CLI is an old way of working with apps and applications and is used to perform specific tasks that users need. CLI is a text-based visual interface, unlike the GUI, which uses graphics options that allow the user to interact with the system and apps. CLI allows the user to perform tasks by entering commands. Its operating system is straightforward but not easy to use. Users enter a command, press "Enter," and wait for a response. After receiving the command, CLI correctly evaluates it and displays the output/effect on the same screen. The command-line interpreter is used for this purpose.

CLI is introduced with a telephone typewriter. This system was based on batch processing. Modern computers support CLI, batch processing, and a single interface GUI. To make good use of CLI, the user must enter a set of commands (one by one) immediately. Many applications (mono-processing systems) still use CLI on their operators. In addition, programming languages like Forth, Python, and BASIC provide CLI. The command-line translator is used to use a text-based interface.

Another feature of CLI is the command line used as a sequence of characters used in the user interface or shell. Command information is used to inform users that CLI is ready to accept orders. MS-DOS is an example of CLI.

Terminal User Interface

TUI is also known as a Text-based user interface. You have text on the screen because they are used only in the terminal. These applications are not well-known to many users, but there are a bunch of them. Terminal-based web browsers are an excellent example of TUI programs. Terminal-based games also fall into this category. Text user interface (also known as written user interaction or terminal user interaction) is a text-based user. TUIs differ from command-line communication in that, like GUIs, they use all of the screen space and do not provide line-by-line output. However, TUIs use only the text and symbols found in the standard text terminal, while GUIs typically use high-definition image terminals.

KDE PLASMA

KDE is more than just a software. A community comprises programmers, contributors, artists, writers, distributors, and users worldwide. The KDE team is committed to building the best free desktop and mobile software. And not only contributors but also users and fans of KDE software can be found worldwide, assisting other users, broadcasting news, or just enjoying the information.

KDE started life as a desktop space 20 years ago. As jobs grow, KDE becomes an international team that creates Free Software and Open Sources. It means making multiple programs from the KDE community work together to provide you with the best computer experience. Does that mean you can't use the KDE program if you don't use the KDE desktop? Not at all. With the help of one or two libraries, applications can be used on almost any Linux desktop. In addition, the software works on a variety of platforms. Now you can find many KDE applications running under Windows and Mac OS or other devices like smartphones and tablets.

Various programs suit users' needs, from simple but powerful text editors to animated audio and video players to an advanced integrated development environment. Also, KDE applications follow a consistent look and feel across the desktop, giving you a comfortable and familiar feeling when using any KDE application.

KDE software has a few other features that make it a great workplace, such as:

- Beautiful and modern desktop.

- The flexible and adjustable program, which allows you to customize applications without having to edit multiple text files.

- Displaying across the network allows you to easily access files on other networks and computers as if they were on your computer.

- Software ecosystem for hundreds, even thousands, of programs.

- Available in more than 60 languages.

Getting KDE Software

If you are using a Windows system or Mac OS, a growing number of KDE applications, such as Krita or Kdenlive, are available for download and installation. You will find installers on their pages.

Some KDE software is for Linux, a free operating system that you can try right now. The community is producing KDE neon, a downloadable Ubuntu Linux operating system that includes a new version of KDE.

KDE HISTORY

KDE, called K(ool) Desktop Environment, was founded in 1996 by Matthias Ettrich, a student at the University of Tübingen. At the time, he was concerned about various aspects of the UNIX desktop. His concern was that not every application looked or behaved the same way. In his view, desktop applications were far more complex for end-users. You have created a desktop environment where users can expect apps to be flexible and easy to use to solve the problem.

The name KDE has been identified as a playground in the existing Common Desktop, which is available on UNIX systems. In contrast, CDE was the X11-based user environment developed by HP, IBM, and the Sun with the X/Open consortium, a visual connector, and a production tool based on the Motif image widget tool kit. It was like being an easy-to-understand desktop computer. K initially represented "Kool," but it was soon decided that K should not represent anything special. KDE was extended to "K Desktop Environment" before being completely replaced by KDE in an attempt to rename.

Initially, Ettrich chose to use the Qt Trolltech Qt framework for the KDE project, but other editors started making KDE/Qt applications, and in 1997, a few applications were released. On July 12, 1998, the first desktop version, KDE 1.0, was released. The original GPL version of the tool kit was only available on forums using the X11 server display. However, with Qt 4, LGPL-licensed versions are available for additional media. It allows QD 4-based KDE software or newer versions to be distributed to Microsoft Windows and OS X at a glance. The KDE team announced the redesign of the KDE project on November 24, 2009. Encouraged by a meaningful change in purpose, the word redesign focuses on highlighting the community of application creators and other tools provided by KDE instead of the desktop.

It was formerly known as KDE 4 but is now split into KDE Plasma Workspaces, Applications, and Platform integrated as KDE Software Compilation 4. Since 2014, KDE now no longer represents the K Desktop Environment but the software community.

Matthias Ettrich first launched the KDE project in 1996. You plan to provide a suitable UNIX-based desktop space for beginner computer users. You have used the GUI, which is more understandable and straightforward for Windows OS users. KDE is currently supported with Linux, Solaris, FreeBSD, OpenBSD, and LinuxPPC. Like LibreOffice in Ubuntu, KDE considered KOffice to be a staple among KDE applications. It includes word processor and spreadsheet, image editing, vector drawing, and presentation applications. The KOffice app was first released in October 2000 as part of the KDE version 2.0 package.

KDE Projects

KDE projects are managed by the KDE community, a group of people who create and promote free software for daily use, for example, KDE Plasma and KDE Frameworks or applications such as Amarok, Krita, or digiKam. There are also non-coding projects such as designing a Breeze desktop and iconset desktop theme, integrated by KDE's VisualDesignGroup. Even non-Qt applications, such as Gcompris, which started as a GTK-based program, or web-based projects like WikiToLearn are an official part of KDE.

KDE neon Linux distribution is based on the long-term Ubuntu release (LTS) integrated with an additional software repository containing the latest 64-bit versions of the desktop Plasma 5 desktop, Plasma 5, Qt 5 framework tools kit, and other KDE compatible software.

You can think of KDE as a GUI for Linux OS. KDE has proven that Linux users make its use easier as they use windows. It provides Linux users a graphical interface to choose their custom desktop location. You can select your graphical interface among the various available GUI areas for their appearance.

You can think of Linux without KDE and GNOME as DOS in windows. KDE and GNOME are similar to Windows, except they are connected to Linux with an x server rather than the operating system. When installing Linux, you select the location of your desktop in two or three different desktop environments such as KDE and GNOME. Another popular site, such as KDE, is GNOME. Both come with other features and some distribution. KDE comes with a variety of features. Some of the key ones are listed next.

Various Parts of the KDE Platform

KDE is a large community of software developers. We all have similarities in building on the infrastructure we have developed over the years: the KDE Platform.

- **KDE Plasma Workspace:** It is a user interface feature optimized for various devices such as PCs, notebooks, or mobile devices. KDE Plasma, a custom desktop architecture with custom layouts and panels, supports virtual desktops and widgets. They are written with Qt 5 and KDE Frameworks 5.

- **KDE Frameworks:** Many KDE applications are built with the help of an integrated framework. It is a collection of software and libraries frameworks built on Qt (formerly known as "kdelibs" or "KDE Platform").

- **KDE Applications:** It is a software program written to use the forum. The applications running such as Kdenlive or Krita are mainly built into KDE Frameworks and are often part of the KDE applications.

KDE CORE PROJECTS

Plasma Workspaces

Workspaces are used to reduce clutter and make the desktop easier to navigate. It also can be used to organize your work. For example, you could have communication windows, such as email and your chat program, on one workspace and your work on a different workspace. Workspaces used in K Desktop Environment 1.1 are given below.

1. KDM

2. KWin

3. Plasma

4. Systemsettings

Workspaces are the term for all graphical environments provided by KDE. Plasma separates components into "data engines" and visualization counterparts. It is intended to reduce the total programming effort when there are multiple possible visualizations of given data and make it easier for the data engine and the workspaces to be written independently. Currently, three workspaces are being developed:

- Plasma Desktop for traditional desktop PCs and notebooks

- Plasma Netbook for netbooks

- Plasma Active for tablet PCs and devices.

There are various workspaces under the plasma.

- **Desktop**: Plasma Desktop is the first workspace that KDE developed. It was declared with the release of KDE SC 4.2. It is designed for desktop PCs and bigger laptops. The default configuration resembles K Desktop Environment 3 and Microsoft Windows XP, but extensive configurability allows radical departures from the default layout. It is a fundamental rewrite of several desktop interaction programs included in earlier KDE desktop environments for UNIX-like systems, focusing on eye candy and special graphical effects. The Desktop Workspace replaces the last KDesktop shell, Kicker taskbar, and SuperKaramba widget engine in the K Desktop Environment 3 with a unified widget system that can replace alternative designs.

- **Netbook**: Plasma Netbook is the second workspace. It aims at netbooks (Netbooks are a category of small, lightweight, legacy-free, and inexpensive laptop computers) and may also be used on tablet PCs – the first stable release shipped with KDE SC 4.4.

- **Plasma Active**: It is not a workspace on its own and a service built on top of the frameworks that enable the creation of full-fledged workspaces using only QML files without the need to program in C++. Plasma Active serves as the base for touch screen-compatible workspaces. Active-compatible releases of the Kontact applications and a document viewer based on Calligra Suite are already available.

- **Contour**: Contour is the Plasma interface for tablet devices. Its development was initiated in April 2011 by basysKom. Replacing an earlier tablet prototype, Contour is now the main workspace and was shipped as part of Plasma Active 1.0, released in October 2011.

- **Mobile**: Plasma Mobiles were targeted as smartphones and small tablet devices mainly used via touch input. They were initially released in 2011 along with Plasma Active 1.0, but the development focus shifted toward Contour. Plasma Mobile, meanwhile, has been superseded by Plasma Active.

Features
- Plasma features essentially an applet that contains other applets. Two primary examples of containments are the desktop background and the taskbar. A containment is anything the developer wants: an image, animation, or even OpenGL. Images are commonly used, but with Plasma, the user could set any applet as the desktop background without losing the applet's functionality. This also allows applets to be dragged between the desktop and the taskbar and different visualization for the more confined taskbar – from KDE 4.0 to 4.2, the default theme such as Oxygen. It was characterized by dark tones. In KDE 4.3, the new Air theme is replaced, which predominates in transparency and white as the base color. New themes for Plasma can be chosen and installed through an authority.

- The Plasma widgets' scalability allows them to be resized and rotated to any size, with only a brief pause to redraw themselves. The Kross scripting framework will enable developers to write widgets in a variety of programming languages in addition to C++. Widgets are their size and can be made to show more or fewer data depending on their size.

- Plasma can support other widgets. SuperKaramba is the widget engine used in the KDE 3 series that has been added for legacy reasons.

KWin – Window Manager

It is a manager for the X Window System. It is an integral part of the KDE Software Compilation, although it can be used independently or with other desktop environments.

There are various window decorations for KWin, including the default Oxygen, Microsoft Windows-like Redmond, and Keramik. IceWM themes can also be used, provided the kdeartwork package is installed. Currently available backends include XRender, OpenGL 1.x, OpenGL 2.x, and OpenGL ES 2.0. KDE 4.3 has the following effects built-in:

Accessibility

- Name
- Invert
- Looking glass
- Magnifier
- Sharpen
- Snap Helper
- Track mouse
- Zoom

Appearance

- Explosion
- Fade
- Fade desktop
- Fall apart

- Highlight Windows
- Login
- Logout
- Magic Lamp
- Minimize animation
- Mouse mark
- Scale In
- Sheet
- Slide
- Sliding pop-ups
- Taskbar Thumbnails
- Thumbnail aside
- Translucency
- Wobbly windows

Window Management

- Box switch
- Cover switch
- Desktop cube
- Desktop Cube Animation
- Desktop Grid
- Flip switch
- Present windows
- Resize window

KHTML – HTML Rendering Engine, Installed on WebKit in 2004

KHTML is a browser engine built by the KDE project. It is the default engine of the Konqueror browser, but it has not been used since 2016. In addition, KHTML will be discontinued in KDE Frameworks 6.

Built on the KParts framework and written in C ++, KHTML has excellent support for Web standards during its first season. The HTML rendering engines are used by some of the world's most widely used browsers, including Google Chrome, Safari, Opera, Vivaldi, and Microsoft Edge. The following levels are supported by KHTML engine.

- HTML 4.01

- HTML 5 support

- CSS 1

- CSS 2.1 (screen and cached media)

- CSS 3 selectors (fully from KDE 3.5.6 [15])

- CSS 3 Other (multiple background, box size and text shadow)

- PNG, MNG, JPEG, GIF image formats

- DOM 1, 2 and section 3

- ECMA-262/JavaScript 1.5

- Scalable Vector Image Support Part

KJS – JavaScript Engine

KJS is a KDE ECMAScript-JavaScript engine developed by the KDE project web browser by Harri Porten in 2000. On June 13, 2002, Maciej Stachowiak announced in the mailing list that Apple was releasing JavaScriptCore, a KJ OS-based Mac OS X framework. With the WebKit project, JavaScriptCore has been transformed into SquirrelFish Extreme, a JavaScript engine that integrates JavaScript into native code.

KIO – Expandable Network File Access

KIO (short for KDE Input/Output) is part of the KDE architecture. It provides access to files, websites, and other resources with a single fixed API. Applications, such as Konqueror, written using this framework, can work on files stored on remote servers in the same way they work on local storage, effectively making the KDE network visible. It allows the file browser as Konqueror, the most flexible and powerful file manager, and web browser. KIO can support individual protocols (e.g., HTTP, FTP, SMB, SSH, FISH, SFTP, SVN, TAR). The KDE Khelp Help Center application has a kioslaves section that lists available terms briefly.

KParts – Frame Part of an Image within a Lightweight Process

KPart technology is used in KDE to re-use GUI components. The advantage presented by KPart is that it comes with predefined toolbar actions. By using KPart in applications, developers can spend less time using a text editor or command-line features, for example, and use katepart or konsolepart instead. KParts is also used with plugin technology to embed applications within one another, such as integrating PIM applications into Kontact.

XML GUI

XML GUI is a KDE framework for designing user interaction applications using XML, using action concepts. In this framework, the editor designs actions that his application can use, with a few actions defined by the editor in the KDE framework, such as opening a file or closing an application. Each action is associated with various data, including thumbnails, captions, and tips.

An exciting part of this design is that the actions are not included in the menus or toolbar by the editor. Instead, the editor provides an XML file explaining the menu bar and toolbar layout. By using this application, the user can rearrange the user interface without having to touch the source code of the application in question.

In addition, XML GUI is useful for the KParts component interface of the KDE component, as the application can easily integrate the KPart GUI into its GUI. Konqueror file manager is a canonical example of this feature. The current version is KDE Frameworks # KXMLGUI.

Phonon – Multimedia Framework

Phonon is a multimedia API provided by KDE and is a standard summary for managing multimedia streaming within KDE software and is used by several Qt applications. Phonon was created to allow KDE and Qt software to be independent of any single multimedia framework such as GStreamer or xine and provide a stable API for the life of the prior version. It is done for various reasons: to build a KDE/Qt-style multimedia API, better support native multimedia frameworks for Windows and macOS, and fix framework issues that may be neglected or have API or ABI instability.

Solid – Device Assembly Frame

Solid is a device integration framework. It provides a query and hardware interface outside of the underlying operating system. It provides the following features for application developers: Hardware Discovery, Power Management, and Network Management.

Sonnet

Sonnet is a spell-checking library based on Qt-based applications plugins. It supports several plugins, including HSpell, Enchant, ASpell, and hunspell. It also supports automatic language detection based on a combination of different algorithms.

ThreadWeaver

It is a system library developed initially for KDE Software Compilation 4 and later redesigned in KDE Frameworks 5. It allows developers to use multi-core processors and multithreading easily. In ThreadWeaver, the workload is divided into individual tasks. Then, there is a relationship between functions. ThreadWeaver will use the most efficient way to use them. Krita used visual filtering preview using ThreadWeaver to disable GUI lock.

KDE APPLICATIONS

Critical applications created by KDE include:

Editors

- **Kate:** Advanced Text Editor for editors
- **Kedit:** A simple text editor with a few features, such as Windows' Notepad
- **Kile:** LaTeX editor
- **KWrite:** Text editor

Education

- **KGeography:** An application that scans local information
- **Kiten:** A Japanese learning tool
- **KLAid:** A learning tool from cards while using a PC
- **Konjue:** A tool for compiling and compiling French verbs
- **Kalzium:** Displays information about the material table

- **KStars:** Planetarium Program
- **Step:** Interactive physics simulation
- **KAlgebra:** Symbolic calculator

Games

- **KAtomic:** Puzzle game
- **KFoulEggs:** Puyo Puyo game
- **Klickety:** Puzzle game
- **KMines:** Minesweeper game
- **Kolf Game:** Golf
- **KReversi:** Othello/Reversi game
- **KSirtet:** Tetris game

Photos

- **DigiKam:** Digital Photography Management
- **Gwenview:** Photo viewer
- **KGhostView:** pdf/ps file viewer
- **Kolourpaint:** A small bitmap photo editor (very similar to MSPaint)
- **KPDF:** PDF viewer
- **KPhotoAlbum:** Digital Photo and Photo Manager
- **KPovModeler:** Modeling and design program for creating POV-Ray scenes
- **Krita:** Bitmap Photo Editor
- **KSnapshot:** Screenshot tool
- **Kuickshow:** Photo viewer
- **KColorEdit:** KDE Color Pallet Editor
- **KView:** Image viewer
- **Okular:** A viewer of international literature
- **ShowImg:** Image viewer

Theme-related Application Groups

- **KDE-Plasma-Addons:** Additional Plasma Widgets

  - KDE-Network

  - KDE-PIM

  - KDE-Graphics

  - KDE-Multimedia

- **KDE Accessibility:** Accessibility applications

  - KDE-Utilities

  - KDE-Edu

- **Calligra Suite:** Integrated office suite

  - KDE-Games

  - KDE-Toys

- **KDE-Art:** Additional icons, styles, etc.

  - KDE-SDK

  - KDE-Bindings

- **KDEWebdev:** It is a web development tool

- **KDE-Extragear:** It is a collection of applications and tools not part of the KDE core applications

- **KDE-Playground:** It contains new and unstable software

Other Projects

- **KDE Connect:** Android application for connecting the Plasma Desktop to phones for remote control

- **KDE Neon:** A distro containing the latest KDE software packages on a Ubuntu basis

- **Wiki2Learn:** A web-based framework for people to participate and share information

System

- **Filelight:** Demonstrates how disk space is used, depicting it as a set of fixed pie charts

- **KBluetooth:** Bluetooth connection

- **KDE Control Center:** A centralized configuration tool

- **KDE System Guard:** Enhanced task manager and system monitoring

- **KDirStat:** Graphical disk utility

- **KDM:** Login Manager

- **Kinfocenter:** Information about your computer

- **KlamAV:** ClamAV antivirus for KDE

- **Konsole:** Terminal emulator

- **KWallet:** Password Manager Protect

- **Yakuake:** A terminal emulator for earthquake style

- **KDE Session Manager:** Session Editor

KDE DEVELOPMENT

Source Code

The source code for all KDE projects is stored in the source code using Git. Stable versions are downloaded to the KDE FTP server in source code with configuration documents. They are ready to be integrated by operating system vendors and integrated with their other systems before distribution. Most users use only stable and tested versions of KDE programs or applications, providing you with easy-to-install, pre-packaged packages.

License

The KDE software project must be released under the accessible license terms. In November 1998, the Qt framework had two licenses under the free and open Q Public License and a commercial license for software developers. In the same year, a KDE Free Qt Foundation was established that guarantees that Qt will fall under the exclusive BSD license if Trolltech ceases to exist or a free version of Qt is released within 12 months.

The debate continued over compliance with the GNU General Public License (GPL), which is why in September 2000, Trolltech made a UNIX version of the Qt libraries available under the GPL over the QPL that removed the concerns of the Free Software Foundation. Trolltech continued to require licenses to develop Qt-related software. KDE's primary libraries have integrated licenses under the GNU LGPL. Still, the only way for the patented software to use would be to develop under the terms of the Qt patent license. Starting with Qt 4.5, Qt was also made available under the LGPL 2.1 version, allowing patented applications to use the Qt version of open-source officially.

Implementation

Many KDE projects use the Qt framework, which works on many applications such as UNIX, macOS, and Microsoft Windows. Since 2011 CMake has been operating as a construction tool. It allows KDE to support a wide range of forums, including Windows. GNU gettext is used for translation. Doxygen is used to produce API documents.

KDE VERSION HISTORY

First, we will discuss software release. There are five versions of KDE.

1. K Desktop Environment 1

2. K Desktop Environment 2

3. K Desktop Environment 3

4. KDE Software Compilation 4

5. KDE Plasma 5

Let's discuss all of the above versions in more detail.

K DESKTOP ENVIRONMENT 1

It was the first release of the K Desktop Environment release. There are two major releases in this series. The development of KDE began shortly after Matthias Ettrich's announcement on October 14, 1996, acquiring the Kool Desktop Environment. Since then, the name Kool has been declining, and the name has just become K Desktop Environment.

Initially, all components were released to the developer community without time covered in the entire project. KDE's first contact with the mailing list was called kde@fiwi02.wiwi.uni-Tubingen.de.

The first release was Beta 1 on October 20, 1997, about one year after the first announcement. Three more beta versions followed on November 23, 1997, February 1, 1998, and April 19, 1998.

K Desktop Environment 1.0

The first version, 1.0 of K Desktop Environments, was released on July 12, 1998. KDE is a desktop platform featuring a network of UNIX workstations. It seeks to fill the need for an easy-to-use desktop for UNIX operating systems, such as desktop applications available under macOS or Windows. The UNIX operating system is the operating system available today. UNIX has been the undisputed choice of information technology for many years. If you look at stability, durability, and openness, UNIX has no competition. However, UNIX's modern, easy-to-use desktop shortages have prevented UNIX from accessing computer user desktops in offices and homes. There is now an easy-to-use, modern UNIX desktop available with KDE. UNIX, such as Linux, UNIX/KDE, is an entirely open computer platform available for free, including source code that must be modified. We hope that the UNIX/KDE combination will eventually bring an open, reliable, stable, and compliant computer to a regular computer.

This version has received mixed approval using the Qt software framework under the Qt Free Edition license, which is incompatible with free software and advises using Motif or LessTif instead. Apart from this, KDE was well received by many users and entered the first Linux stream.

K Desktop Environment 1.1

The K Desktop Environment 1.1 version was faster, more stable, and included minor improvements. It had a new set of images, backgrounds, and styles. Some sections have received additional updates, such as Konqueror pre-kfm, kpanel app launcher, and KWin pre-kwm. Then, recently introduced e.g., kab, the address management library, and KMail rewriting, called kmail2, were installed as alpha in line with the older version of KMail. Kmail2 did not leave the alpha status, and the upgrade was completed in the old KMail update. K Desktop Environment 1.1 has been well received by critics. At the same time, Trolltech prepared version 2.0 for Qt, which was released as a beta in 1999-01-28. Next, no more QD 1-based KDE 1 development is underway. Instead, only bug fixes were released: version 1.1.1 in 1999-05-03 and version 1.1.2 in 1999-09-13. The most in-depth upgrade and port to Qt 2 were upgraded as K Desktop Environment 2.

K DESKTOP ENVIRONMENT 2

It was the second release of K Desktop Environment (now KDE Software Compilation). There were two major releases in the series.

Big Updates

1. K Desktop Environment 2 introduced significant technological improvements compared to its predecessors.

2. DCOP (Desktop Communication Protocol) is a client–client communication protocol connected to a server over a standard X11 ICE library. Its goal for the system was to allow applications to interoperate and share complex tasks. DCOP was a remote control system that allowed applications or scripts to enlist the help of other applications. It was built on top of the X Window System Inter-Client Exchange protocol.

   It provides extensive new capabilities without requiring entirely new applications to be written. KDE applications and the KDE libraries did make heavy use of DCOP. Most of the KDE applications can be controlled by scripts via the DCOP mechanism. D-Bus replaced DCOP in KDE Software Compilation 4. A command-line tool called "dcop" can be used for communication with the applications from the shell, where "kdcop" is a GUI tool to explore the interfaces of an application.

DCOP Model

The DCOP model is simple. Each application using DCOP is a client. They communicate through a DCOP server, which functions like a traffic director, dispatching messages/calls to the proper destinations. All clients are peers of each other. Two types of actions are likely with DCOP: "send and forget" messages, which do not block, and "calls," which block waiting for some data to be returned.

Any data sent is serialized (also referred to as marshalling in CORBA speak) using the built-in QDataStream operators available in all Qt classes. A simple IDL-like compiler is available – dcopidl or dcopidl2cpp – that generates stubs and skeletons using the dcopidl compiler benefits safety. The Trinity Desktop Environment uses it.

1. KIO, I/O library. The network is transparent, which can access HTTP, FTP, POP, IMAP, NFS, SMB, LDAP, and local files. In addition, it allows developers to "enter" some of the basic features, such as WebDAV, which are automatically available in all KDE applications. It can also find holders for certain types of MIME.

2. These handlers can be embedded within the application using KParts technology.

3. KParts is a component object model which allows an application to embed another within it. When activated, the component handles all embedding features, such as toolbar settings and relevant menus. It can communicate with KIO to find brokers for specific MIME types or services/protocols.

4. KHTML, an HTML 4.0 listening engine for rendering and drawing. It supports many JavaScript, Java, HTML 4.0, CSS 2, and SSL for secure communication. Compatible with Netscape plugins like Flash. KHTML can also embed components within it using KParts technology.

Default Environment

These operating systems offer it as the default environment:

- ALT Linux
- Ark Linux
- ArtistX
- aptosid
- BackTrack
- Chakra Linux
- Frugalware
- Kanotix
- Kororaa

- Kubuntu

- Mageia (DVD version)

- Mandriva Linux

- Magic Linux

- MCNLive

- MEPIS

- Netrunner (operating system)

- openSUSE

- Pardus

- PCLinuxOS

- Qomo Linux

- Sabayon Linux

- Slackware

- Skolelinux

- VectorLinux

- Z-Soft

- YOPER

- PC-BSD

- BeleniX

- SuperX

System Settings

It is a KDE application used to configure the system under KDE Plasma Workspaces. It replaces K Desktop Environment 3's KControl.

Features

- Control Center for global KDE Platform settings

- All desktop settings converged on one central location

- General and Advanced tabs separate most common user settings
- The search function helps narrow down probable settings

K Desktop Environment 2.0

Konqueror is introduced as a web browser, file manager, and document viewer. Use KHTML to display web pages. K Desktop Environment 2 is also the first release of the KOffice suite, which includes a spreadsheet program like KSpread, a vector drawing application like KIllustrator, a word-based word processing app like KWord, a KPresenter-like presentation program, and a chart and drawing app like KChart. Indigenous file formats are based on XML. The KOffice suite can encrypt the components into KParts. The following words are used in the above lines.

Konqueror

A web browser and file manager provide file viewer functionality to systems such as remote FTP server files, local files, and disk image files. It is developed by developers and can work on many operating systems such as UNIX and Windows. It is an integral part of KDE Software Compilation. It is licensed and distributed under version 2 of the GNU General Public License.

The term "Konqueror" referred to two main competitors when the Konqueror browser was released for the first time: "first comes Navigator, Explorer, and then Konqueror." It follows the KDE naming convention, and most KDE programs start with the letter K.

It was released in version 2 on October 23, 2000. It replaces the previous version, KFM (KDE file manager). With the release of KDE4, Konqueror was replaced as Dolphin's default file manager.

Largely Supported Protocols

- FTP Browser and SFTP/SSH
- SAMBA Browser (Microsoft file sharing)
- HTTP Browser
- IMAP mail client
- ISO Viewer (CD)
- VNC Viewer

Konqueror User Interface

Konqueror's user interface for Microsoft's Internet Explorer can be customized. Works well with "panels" that can be edited or added. For example, the online bookmarks panel on the left side of the browser and the appropriate web page will be viewed in the main right panel by clicking the bookmark. Alternatively, list the categories of folders in one panel and the content of the selected folder in another. The panels are flexible and can include, among other KParts, components, a console window, a text editor, and a media player. Panel settings can be saved, and there is an automatic configuration.

Navigation functions, such as background, front, history, etc., are available during all operations. Most keyboard shortcuts can re-map using graphical configuration, and navigation can be done by assigning characters to nodes in the active file by pressing the control button. The address bar has auto-complete location references, past URLs, and past search terms.

The app uses a tab-based interface, where the window can contain multiple texts in the tabs. Many virtual connectors are not supported; however, it is possible to repeatedly split a window to view multiple documents at once or open another window.

Konqueror Web Browser

Konqueror is built as a stand-alone web browser project. It uses KHTML as its structural engine, is compliant with HTML, and supports JavaScript, Java applets, CSS, SSL, and other appropriate open standards. Another building engine, bkitpart, is available in Extragear.

While KHTML is the default search engine provided on the web, Konqueror is a modular application, and other rendering engines are available. In particular, WebKitPart using the WebKit engine derived from KHTML has greatly supported the KDE 4 series.

K Desktop Environment 2.1

Using a modular plugin design, K Desktop Environment 2.1 has released a noatun media player. For upgrades, K Desktop Environment 2.1 is integrated with KDevelop.

K Desktop Environment 2.2

The release of KDE 2.2 has improved the start-up time of applications in GNU/Linux programs and increased HTML rendering and

JavaScript stability and capabilities. Several new plugins have been added to Konqueror. KMail received IMAP support, including SSL and TLS, while KOrganizer received native iCalendar support. Other improvements include a new plugin-based print architecture and a personalization wizard.

K DESKTOP ENVIRONMENT 3

The third release of K Desktop Environment is now called KDE Software Compilation. There are six major versions in the series.

K Desktop Environment 3.0

Introduce limited usage support, feature-specific features such as products, online cafes, and business feeds, which do not allow full access to all software piece capabilities to prevent specific undesirable actions. KDE version 3.0 has introduced a new lock framework to address these needs, especially the system-based configuration-based system that complements the UNIX basic system permissions. KDE panel and desktop administrator have been modified to use the method. However, other significant desktop components, such as Konqueror and Control Center, had to wait for the release.

K Desktop Environment 3.0 has a new printing framework, KDEPrint. KDEPrint module design supports various print engines, such as CUPS, LPRng, and LDP/LPR. In conjunction with CUPS, KDEPrint is managed. KDEPrint provides a command-line interface, including GUI configuration features access to non-KDE programs, such as OpenOffice.org, Mozilla Suite, and Acrobat Reader.

This release has introduced the KDE address book library with the central address book for all KDE applications. The library is based on the vCard standard and expanded with additional backgrounds such as LDAP and server servers.

K Desktop Environment 3.1

K Desktop Environment 3.1 with Konqueror and the About screen. Version 3.1 has introduced a new automatic window (Ceramic), icon styles (Crystal), and enhancements. The update included an enhanced LDAP integration with KMail's enhanced Kontact security. It also supports S/MIME, PGP/MIME, and X.509v3. The desktop framework presented in version 3.0 was expanded. Other improvements include:

- It is used to interrupt browsing in Konqueror.

- KGet is a new download manager.

- New Xine-based multimedia plugin and desktop sharing framework.

K Desktop Environment 3.2

K Desktop Environment is version 3.2 with Konqueror and the About screen. This release is described in the K Desktop Environment.

K Desktop Environment version 3.2 includes new features such as checking spelling forms and emails, tabs in the Konqueror program, enhanced email support, and the Microsoft Windows desktop sharing protocol (RDP). Freedesktop.org's performance and standards have been improved by lowering operating times and strengthening interoperability with other Linux and UNIX software. After the KDE community worked with Apple Safari's web browser team, KDE's web support functionality increased and compliance with web standards.

KDE desktop environment has improved usability with multiple applications, dialogs, and control panels to focus on clarity, usability, and minimizing clutter in various menus and toolbars. Hundreds of new icons are being created to enhance the natural harmony, with changes in the default viewing style, including new screens, vibrant progress bars, and stylish panels. The plastic style has been removed from this release.

NEW APPLICATIONS INSTALLED

JukeBox

It was started by Scott Wheeler in 2000 and was initially called QTagger and was the first official part of KDE in KDE 3.2. JuK, a jukebox music player. KDE Free Software Audio Player, a default player from K Desktop Environment 3.2. and supports MP3, Ogg Vorbis, and FLAC audio files. It is an audio jukebox app that focuses on music management.

It has the following features.

- Collection and playlist defined by most users.

- Able to scan directories to automatically import playlists (files.m3u) and music files initially.

- Powerful Search Playlists can be updated automatically as group fields change.

- It has a tree viewing mode, where playlists automatically record albums, artists, and genres.

- Has a playlist history showing which files were played and when.

- Support online search to filter the list of objects.

- Can predict tag information in the file name.

- File renaming can rename files based on tag content.

- Support reading and editing tags ID3v1, ID3v2, and Ogg Vorbis (via TagLib).

Kopete

Kopete is a multi-protocol, and free messaging client software downloaded from KDE Software Compilation. However, it can work in many areas, designed and integrated with KDE Plasma Workspaces. A fan nominated by KDE Telepathy from the KDE RTCC Initiative.

Features

- Allows collecting messages within a window, with easy-to-change chats

- It can use multiple accounts for many services

- Contact nicknames

- Allow contacts

- Custom contact notifications

- Integration of KAddressBook and KMail

- Chat login

- Contains custom emoticons

- Custom notifications, including pop-ups and sounds

- QQ and Yahoo! messenger webcam support

- On the fly spell checking

- Voice calls via GoogleTalk and Skype

KWallet

KWallet (short for KDE Wallet Manager) is a desktop management system for the KDE Software Compilation. Provides a way to store sensitive password encrypted files, called "wallets." For added security, each wallet can be used to store different types of authentication, each with its password.

Kontact

Kontact is the information and software manager of the groupware group developed by KDE. Supports contacts, calendars, notes, to-do lists, news, and email. Provides many flexible image UI, such as KMail, KAddressBook, Akregator, etc., built on a common theme.

Other KDE Applications

- KIG, interactive geometry system

- KSVG, SVG viewer

- KMag, KMouseTool, and KMouth are new access tools.

- KGoldRunner, a new game

K Desktop Environment 3.3

K Desktop Environment version 3.3 focuses on contrast. K Desktop Environment version 3.3 focused on different desktop components. Kontact was combined with Kolab, a groupware application, and Kpilot. Konqueror is given better support for instant messaging contacts, can send files to IM contacts, and supports IM protocols. KMail has shown the ability to display the online presence of IM contacts. Juk support for burning audio CDs with K3b.

This update also included many minor desktop enhancements. Konqueror application received tab improvements, an RSS feed viewer sidebar, and a search bar compatible with all keyword searches. KMail was given HTML composition, anti-spam, antivirus wizards, automatic handling of mailing lists, support for cryptography, and a quick search bar.

New Applications Included

- Kolourpaint is a KPaint replacement.

- KWordQuiz, KLatin, and KTurtle are education packages for schools and families.

- Kimagemapeditor and klinkstatus are tools for web designers.

- KSpell2, a new spell-checking library, is improving on KSpell's shortcomings.

- KThemeManager is a new control center module for the global handling of KDE visual themes.

K Desktop Environment 3.4

K Desktop Environment version 3.4 focused on improving accessibility. The update added a text-to-speech system supporting Konqueror, Kate, KPDF, the standalone application KSayIt, and text-to-speech synthesis on the desktop. A high contrast style, a complete monochrome icon set was added, and an icon effect of painting all KDE icons into any two arbitrary colors.

Kontact support for various groupware servers, whereas Kopete was integrated into Kontact. KMail can store passwords securely in KWallet. KPDF can select and copy and paste text and images from PDFs, along with other improvements. The latest update added a new application, Akregator, which can read news from various RSS-enabled websites all in one application.

The latest update added D-Bus/HAL support to allow dynamic device icons to keep in sync with the state of all devices. The kicker has improved visual aesthetic, and the trash system was redesigned to be more flexible. The new desktop allows SVG to be used as wallpapers. KHTML was enhanced standards support, having nearly complete support for CSS 2.1 and CSS 3. KHTML plugins are allowed to be activated on a case-by-case basis. There are also improvements to the way Netscape plugins are handled.

K Desktop Environment 3.5

The K Desktop Environment 3.5 has SuperKaramba, which provides integrated and simple-to-install widgets to the desktop. Konqueror's ad-block feature made the second web browser pass the Acid2 CSS test, ahead of Firefox and Internet Explorer. Kopete application gained webcam support for the MSN and Yahoo! IM protocols. The edutainment module has three new applications, KGeography, Kanagram, and blinKen. Kalzium also saw improvements.

The Trinity Desktop Environment

The Trinity Desktop Environment project, organized and led by Timothy Pearson, Kubuntu release manager for KDE 3.5, has released Trinity to pick up where the KDE e.V. left. This keeps the KDE 3.5 branch alive and releases bugs fixes, additional features, and compatibility with recent hardware. Trinity is packaged for Debian, Ubuntu, Red Hat, and other distributions.

The Kolab Enterprise packages are still being developed and tested on Kontact 3.5. A version based on Kontact 4 is available but not yet recommended for regular use.

KDE SOFTWARE COMPILATION 4 (KDE SC 4)

KDE Software Compilation 4 is the current series of releases of KDE Software Compilation. The first major version (4.0) of the series was released on January 11, 2008, and the latest major version (4.10) was released on February 6, 2013. Whereas major releases (4. x) come within six months, minor bugfix releases (4.x.y) come monthly.

The new series includes several of the KDE Platform components, which contains an API called Phonon with framework Solid having a default icon called Oxygen. It has a new unique desktop and panel user interface called Plasma, which supports desktop widgets, replacing K Desktop Environment 3's separate components.

KDE Platform 4 makes it easy for KDE applications to be portable to different operating systems and made possible by the port to Qt version 4, which supports non-X11-based platforms having Microsoft Windows and Mac OS X. Versions 4.0 to 4.3 of the KDE Compilation known simply as KDE 4. The change was a component of the KDE project's rebranding to reflect its increased scope. There are various released versions of KDE 4 as follow,

KDE 4.0

Most development is implemented in most of the new technologies and frameworks of KDE 4. Both Plasma and the Oxygen style are two of the most significant user-facing changes.

Now, Dolphin replaces Konqueror as the default file manager in KDE 4.0. It addresses complaints of Konqueror being complicated for a simple file manager. However, Dolphin and Konqueror share as much code as possible, and Dolphin can embed it in Konqueror to allow Konqueror to be still used as a file manager.

Okular replaces other document viewers used in KDE 3, like KPDF, KGhostView, and KDVI. It uses software libraries and can be extended to view almost any document. Like Konqueror and KPDF in KDE version 3, Okular can also be embedded in other applications.

Release

On January 11, 2008, KDE 4 was released. Despite being labeled a stable release, it was intended for early adopters. Using KDE 3.5 was suggested for users wanting a more stable, "feature complete" desktop.

The release of KDE 4.0 met with a mixed reception. While early adopters tolerated the lack of finish for some of its new features, the release was widely criticized for its lack of stability and its "beta" quality.

Computerworld reporter Steven Vaughan-Nichols criticized KDE 4.0 and 4.1 and called for a fork of KDE 3.5 by rebuilding on top of Qt 4. The same reporter praised KDE 4.3 and welcomed Trinity's KDE 3.5 continuation project. However, Linus Torvalds switched from GNOME to KDE in December 2005, GNOME after Fedora replaced KDE 3.5 with 4.0. In an interview with Computer World, he described KDE 4.0 as a "break everything" model and "half-baked" release, claiming that he expected it to upgrade KDE version 3.5. Significant features were being regressed due to its extensive changes.

Major Updates of KDE 4

Many applications such as Extragear and KOffice modules have acquired numerous improvements with the new features of KDE 4 and Qt 4. But since they follow their release schedule, they were not available at the time of the first KDE 4 release. These include applications such as Amarok, K3b, digiKam, KWord, and Krita. The Qt 4 series enabled KDE 4 to use less memory and be noticeably faster than KDE 3. The KDE libraries have been made more efficient. However, KDE 4.4 has the highest memory utilization on default Ubuntu installations than GNOME 2.29 Xfce 4.6.

The version LXDE 0.5. Qt version 4 is available under the LGPL for Mac OS X and Windows operating system, which allows KDE version 4 to run on those platforms. In August 2010, KDE Software Compilation four on Mac OS X was considered beta, while Windows is not in the finishing state so that applications can be unsuitable for day-to-day use. Both ports use as little divergent code as possible to make the applications function almost identically on all platforms. In Summer of Code, an icon cache was created to decrease application start-up times in KDE 4. Improvements were varied – Kfind, an application that used several hundred icons. Other applications and an entire KDE session started a little over a second faster.

Pre-releases

KDE 4.0 Alpha 1 was released in the market after adding significant features to KDE base libraries, shifting the focus onto integrating the new technologies into applications and the primary desktop. Alpha 1 had new frameworks to build applications with, providing improved hardware and multimedia integration through Solid and Phonon. Dolphin and Okular were integrated, and a unique visual appearance was provided through Oxygen icons.

Alpha 2 was released mainly focused on integrating the Plasma Desktop, improving the functionality, and stabilizing KDE.

Beta 1 was released with significant features included a pixmap cache – KDE PIM improvements, speeding up icon loading, improved KWin effects, and configuration, better interaction between Konqueror and Dolphin, and Metalink support added KGet for enhanced downloads.

- Beta 2 was released with the support of BSD and Solaris. The release included:

- The addition of the Blitz graphic library.

- Allowing developers to use high-performance graphical tricks like icon animation.

- KRDC (K Remote Desktop Client) overhaul for Google's Summer of Code.

- Plasma provides Amarok's central context view.

Beta 3 was released release was focused on stabilizing finishing the design of libraries for the release of the KDE Development Platform. Plasma had many new features, including an applet browser. The educational software received many improvements like Marble and Parley with bug fixes in other applications.

Beta 4 was released. A list of release blockers was compiled, listing issues that need to be resolved before KDE starts with the desktop's release candidate cycle. The goal is to focus on stabilization and fixing the release blockers. At the same time, the first release of the KDE 4.0 Development Platform was released containing all the base libraries to develop KDE applications, including "widget libraries, a network abstraction layer, various libraries for multimedia integration, hardware integration to resources on the network."

Let's discuss the version of KDE 4

KDE 4.1

It was released on July 29, 2008, and included a shared emoticon theming system used in PIM, Kopete, and DXS, which lets applications download and install data from the Internet with one click. Also introduced are the GStreamer application, QuickTime 7, and DirectShow 9 Phonon backends. Plasma improvements support Qt 4 widgets and WebKit integration, allowing many Apple Dashboard widgets to be displayed. There are also be ports of some applications to Windows and Mac OS X.

New applications include:

- Dragon Player multimedia player (formerly Codeine)
- Kontact with some Akonadi functionality
- Skanlite is a scanner application
- Step physics is a simulator
- Games – Kdiamond Kollision, Kubrick, KsirK, KBreakout

KDE 4.2

KDE 4.2 shows KMail, Dolphin, and was released on January 27, 2009. The release is viewed as a significant improvement beyond KDE 4.1 in all aspects and with a suitable replacement for KDE 3.5 for most users.

KDE Workspace Improvements

The 4.2 release includes the number of bug fixes that have implemented many features present in KDE 3.5 had missing in KDE 4.0 and 4.1. These include multiple row layout and grouping in the taskbar, icon hiding in the system tray, panel auto-hiding, window previews tooltips are back in the panel and taskbar, notifications, job tracking by Plasma, and have icons on the desktop using a Folder View as the desktop background where icons remain where they are placed.

New Plasma includes leaving messages on a locked screen, previewing files, switching desktop activities, monitoring news feeds, utilities such as the Pastebin applet, the calendar, timer, unique character selector, a QuickLaunch widget, and a system monitor, among many others. The Plasma workspace can load Google Gadgets. Its widgets can be written in Ruby and Python. Also, support for applets written in JavaScript and Mac OS X widgets has been improved. New desktop alterations have been added, such as the Magic Lamp, Minimize impact, and the Cube and Sphere desktop switchers. Other modifications, such as the Desktop Grid, have been improved. The user can easily choose effects that have been reworked to select the most commonly used results. Compositing desktops have been enabled by default, where drivers can support them. It can automatic checks confirm that compositing works before allowing it on the workspace.

KRunner has extended functionality via several new plugins, including spell-checking, Konqueror can access browser history, control power through PowerDevil, KDE Places, Documents, and the ability to start specific sessions of the Kate editor, Konqueror, and Konsole. The converter plugin also supports quickly converting between speed, mass, and distance units. Multi-screen support has been improved via the Kephal library by fixing multiple bugs when running KDE on more than one monitor.

KDE 4.3

KDE 4.3 desktop was released on August 4, 2009, showing Dolphin, KMail, and a selection of desktop widgets, with this release being described as incremental and lacking in significant new features. It fixed over 10,000 bugs and implemented almost 2,000 feature requests. Also, integration with other technologies, such as PolicyKit, NetworkManager and Geolocation services, was another focus of this release. A more flexible system tray has developed many new Plasmoids, including the openDesktop.org plasmoid, the first take on the Social Desktop. Plasma also receives more keyboard shortcuts.

KDE SC 4.4

It was released on February 9, 2010, based on version 4.6 of the Qt 4 toolkit. As such, KDE SC 4.4 has Qt's performance improvements and Qt 4.6's new features, such as the new animation framework Kinetic. A completely new application replaces KAddressBook with the same name – previously tentatively called KContactManager.The new KAddressBook is Akonadi integration and has a streamlined user interface. Another significant new feature is an additional new Plasma interface targeted toward netbooks. Kopete is released as version 1.0. KAuth, a cross-platform authentication API, is made in KDE SC 4.4. Initially, only PolicyKit is supported as a backend.

KDE SC 4.5

KDE SC 4.5 was released on August 10, 2010. New features include integrating the WebKit library, an open-source web browser engine used in major browsers such as Apple Safari and Google Chrome. KDE's KHTML engine will continue to be developed, whereas KPackage has been deprecated, and KPackageKit was suggested to replace it, but it didn't make it replace it.

KDE SC 4.6

KDE SC 4.6 was released on January 26, 2011, and had better OpenGL compositing along with a myriad of fixes and features.

KDE SC 4.7

It was released on July 28, 2011. The version updated KWin to be compatible with OpenGL ES 2.0, which will enhance its portability to mobile and tablet platforms. Other optimizations, such as Qt Quick, were made to strengthen this portability. This version brought some updates and enhancements to Plasma Desktop, such as better network management and updates to certain widgets and activities.

Apart from the desktop environment, version 4.7 updates many applications within the Software Compilation. The Dolphin file manager has been updated to provide a clean user interface. Now it supports voice navigation, map creation, and new plugins. The Gwenview image viewer allows users to compare more than two photos side by side. The Kontact database has been ported to Akonadi, allowing the database to be accessible from other applications.

DigiKam has been supporting face detection, image versioning, image tagging. Most of the applications such as Kate, Kalzium, KAlgebra, KStars, and KDevelop have been updated. Moreover, version 4.7 fixed over 12,000+ bugs.

KDE SC 4.8

KDE SC Release 4.8 was available on January 25, 2012.

Plasma Workspaces
KWin performance was increased by optimizing effect rendering. Window resizing was improved as well. Another KWin is QML based Window switcher, initial Wayland support and AnimationEffect class.

Applications
A new version of Dolphin shipped with KDE Applications 4.8. It has improved performance with better file animated transitions name display with other new and improved features.

KDE SC 4.9

KDE SC 4.9 was available on August 1, 2012. The release featured various improvements to the Dolphin file manager, including the reintroduction of in-line file renaming, back and forward mouse buttons, the advance of the places panel, and better usage of file metadata. More, there were several improvements to KWin and Konsole. Activities were better integrated with the workspace—several updated applications, including Okular, Kopete, Kontact, and educational applications.

KDE SC 4.10

It was released on February 6, 2013. Many default Plasma widgets were rewritten in QML, and Nepomuk, Kontact, and Okular improved significantly.

KDE SC 4.11

KDE SC 4.11 was released on August 14, 2013. Kontact and Nepomuk received many optimizations. The first generation Plasma Workspaces entered maintenance-only development mode.

KDE SC 4.12

KDE SC 4.12 was launched on December 18, 2013. The Kontact received substantial improvements.

KDE SC 4.13

KDE SC 4.13 was launched on April 16, 2014. The Nepomuk semantic desktop search was replaced with KDE's in-house Baloo. KDE SC 4.13 was released in 53 different translations.

KDE SC 4.14

KDE SC 4.14 was launched on August 20, 2014. The release primarily focused on stability, with numerous bugs fixed and a few new features added. It was the final KDE SC 4 release.

KDE PLASMA 5

It is the fifth and current generation of the visual workspace environment created by KDE, mainly for Linux systems. KDE Plasma version 5 is the successor of KDE Plasma 4 and was released on July 15, 2014.

It includes a new theme, known as "Breeze," and increased convergence across different devices. The graphical interface was migrated to QML, which uses OpenGL for hardware acceleration, which resulted in better performance and reduced power consumption. Its Mobile is a Plasma 5 variant for Linux-based smartphones.

Software Architecture

KDE Plasma version 5 is built using Qt 5 and KDE Frameworks 5. It improves support for HiDPI displays and a convertible graphical shell, adjusting. KDE 5.0 also includes a new default theme. Qt 5's QtQuick 2 uses a hardware-accelerated OpenGL scene graph to compose and render graphics on the screen, allowing the offloading of computationally expensive graphics rendering tasks onto the GPU, freeing up resources on the system's main CPU.

It uses the X Window System. It supports that Wayland was prepared in the compositor and planned for a later release. It was created initially available in the 5.4 release. Stable support for an introductory Wayland session was provided in the 5.5 release (December 2015).

Support for NVIDIA proprietary driver for Plasma on Wayland was added in the 5.16 release (June 2019).

Development

Since the KDE Software Compilation split into KDE Plasma, KDE Frameworks, and KDE applications, each subproject can develop at its own pace. KDE Plasma 5 is on its release schedule, with feature releases every four months and bugfix releases in the intervening months.

Workspaces

The latest Plasma 5 features the following workspaces:

- Plasma Desktop for any mouse or keyboard driven computing devices like desktops or laptops

- Plasma Mobile for smartphones

- Plasma Bigscreen for TVs and set-top boxes incl. voice interaction

- Plasma Nano, a minimal shell for embedded and touch-enabled devices, like IoT or automotive

Desktop Features

- KRunner is a search feature with many available plugins. In addition to launching apps, can find files and folders, open websites, convert from one currency to another, calculate simple mathematical expressions, and perform numerous other valuable tasks.

- Flexible desktop and panel layouts composed of Widgets, also known as "Plasmoids," can be configured, moved around, replaced with alternatives, or deleted. Each screen layout can be individually configured. New widgets created can be downloaded within Plasma.

- Have a powerful clipboard with a memory of pieces of text that can call up at will.

- Systemwide notification system supporting fast reply, drag-and-drop straight from notifications, history view, and a Do Not Disturb mode.

- Central location to control media playback in open apps, your phone, or your web browser.

- Activities allow you to separate methods of using the system into distinct workspaces. Each activity can have a set of favorite and recently used applications, wallpapers, "virtual desktops," panels, window styles, and layout configurations. It also couples with ksmserver (i.e., X Session Manager implementation), which keeps track of apps that can be run or shut down along with given activity via subSessions functionality that keeps track of applications (not all applications support this feature as they don't implement XSMP protocol).

- Encrypted vaults for storing sensitive data.

- Night Color can automatically warm the screen colors at night, user-specified times, or manually.

- Style icons, cursors, application colors, user interface elements, splash screens, and more can change. Global Themes allow the entire look of the system to be modified in one click.

- Session Management allows apps running when the system shuts down to be automatically restarted in the same state they were in before.

LINUX DISTRIBUTIONS USING PLASMA

Plasma 5 is a default desktop environment on Linux distributions, similar to:

- ArcoLinux
- Fedora – KDE Plasma Desktop Edition is an official Fedora spin distributed by the project
- KaOS
- KDE neon
- Kubuntu
- LliureX
- Manjaro – as Manjaro KDE edition
- MX Linux
- Netrunner
- openSUSE
- PCLinuxOS
- Q4OS
- Slackware
- Solus Plasma
- SteamOS 3.0
- Ubuntu Studio

History

The first Technology Preview of Plasma 5 was released on December 13, 2013. The first release version, Plasma 5.0, was on July 15, 2014. In 2015, Plasma 5 replaced Plasma 4 in many popular distributions, such as Fedora 22, Kubuntu 15.04, and OpenSUSE Tumbleweed.

Releases

Feature releases are released every four months and bugfix releases in the intervening months. Following version 5.8 LTS, KDE plans to support each new LTS version for 18 months with bug fixes, while new regular releases will see feature improvements. Here is the complete list of the KDE release.

KDE 5.0

July 15, 2014: KDE announces the immediate availability of Plasma 5.0, providing an updated and user-friendly desktop experience. Plasma 5.0 introduces a large new version of the KDE workspace. The new concept of Breeze art introduces pure visuals and improved readability. The central workflow is simplified, while the well-known interaction patterns are left as is. Plasma 5.0 enhances high DPI display support and delivers a flexible shell, which is able to switch between user experiences on different targeted devices. Changes under the cover include a move to a new, fully accelerated hardware stack based on OpenGL (ES) scenegraph. Plasma is built using Qt 5 and Frameworks 5.

Significant changes to this new version include:

- It is an updated and modern user, visually clean and interactive where the new Breeze theme is very different, a flat workplace theme.

- It is available in a variety of light and dark. Lightweight and simple graphics and typography-based layouts provide a clean and clear user experience.

- Smooth photo functionality thanks to the updated photo stack

- The Plasma user interface is provided at the top of the OpenGL or OpenGL ES scenegraph, loading multiple computer-enabled delivery functions.

- It allows for higher frames and smoother image displays while freeing up core system processor resources.

KDE Plasma 5.1

October 15, 2014: KDE releases Plasma 5.1.0, the first release featuring new features since the release of Plasma 5.0 in the summer of 2014. Plasma Games 5.1 is a wide variety of developments, leading to more excellent stability, better performance, and new and improved features. Thanks to public feedback, KDE developers were able to pack a large number of fixes and enhancements for this release, among which is a complete and high-quality artwork that follows the new 5.0 Breeze style, a re-addition of popular features like these. such as the Icon Tasks task switch and improved stability and performance.

Regular travelers will enjoy better time-based support on the panel clock. At the same time, those who stay at home have an updated clipboard manager, allowing you to access past clipboard content easily. The Breeze widget is now also available in Qt4-based applications, leading to greater interoperability across all applications. Wayland's support function as a Plasma display server is ongoing, with enhanced support, but not limited to 5.1. Changes to all default sections improve access for visually impaired users by adding screen reader support and enhanced keyboard navigation.

In addition to visual enhancement and features, this release focuses on stability and performance enhancement, with more than 180 bugs resolved from 5.0 in the shell alone. KDE Plasma 5.1 requires KDE Frameworks 5.3, which brings a more significant number of improvements and performance improvements than a large number of adjustments to Plasma 5.1. If you want to help make most of this happen, consider a donation to KDE so that we can support more developers coming together to create great software

Changes Made in the Plasma 5.1 Version
Visual

The new Breeze Qt 4 widget theme allows KDE Platform 4 to fit your Plasma 5 desktop. The concept of Breeze art, which first appeared in Plasma 5.0, has seen many improvements. The icon set is now quite complete. Instead of the notifications on the panel, the icons are visually touched. A new style of native widget enhances the delivery of applications used in Plasma. The unique native style also works for Qt 4, allowing applications written with KDE Platform 4 to fit your Plasma 5 desktop. A new Settings module will enable you to switch between desktop themes.

Overall, Plasma 5.1's Look and Feel improved the information available in 5.0 significantly. Behind all these changes is the development of the Human Interface Guidelines, which have resulted in consistent user information.

New and Old Features

It brings many features that users have grown accustomed to since its 4.x predecessors. Additional widgets are famous as the Token Activity

Manager only. The Notes widget and the System Load viewer are re-entered. Multi-location support has been added to the panel clock. Notifications have been significantly improved and fixes for many significant and minor bugs.

The new feature allows you to switch between different widgets sharing the same purpose easily. Changing the app launcher as an example is very easy to find. Plasma panels have new switches for easy switching between different widgets with the same functionality. You can easily choose which app menu, clock, or task manager you want. The new clipboard widget provides a redesigned user interface over the respected Plasma clipboard manager, allowing the user to easily use the clipboard history and preview files currently on the clipboard. Another Plasma launcher, Kicker, has seen tremendous development, including better accessibility and integration with the package manager.

Wayland

The KWayland Library provides information on setting up Wayland at KInfoCenter and other users. More and more work is needed to run the Plasma operating system in Wayland.

Eligibility and Updates

Plasma 5.1 provides a desktop with a feature set that will suffice for most users. The development team focuses on the tools that make up the central workflow. Although many known features in the Plasma 4. x series are already available in Plasma 5.1, not all of them are installed and made available in Plasma 5 at the moment. As with any software release of this size, bugs may make migration to Plasma 5 difficult for some users. The developers would like to hear about any problems you may encounter to be resolved and resolved. We have made a list of the issues we know and work on. Users can expect monthly bugfix updates. Releases that bring new features and bring back even older features will be made in early 2015.

KDE Plasma 5.2 Release

January 27, 2015: KDE releases Plasmas 5.2. It added several new features and many bug fixes.

This Plasma release comes with new features to make your desktop even more complete:

- **BlueDevil:** a list of desktop components to manage Bluetooth devices. It will set up your mouse keyboard, send and receive files, and you can browse devices.

- **KSSHAskPass:** if you access computers with ssh keys, but those keys have passwords, this module will give you an image UI to enter those passwords.

- **Muon:** It installs and manages software and other addons on your computer.

- **Login theme setting (SDDM):** SDDM is now the default Plasma login manager, and this new System Settings module allows you to customize the theme.

- **KScreen:** detecting its first release of Plasma 5 is a System Settings module to set up multiple monitoring support.

- **GTK Program Style:** this new module allows you to customize the theme of applications from GNOME.

- **KDecoration:** The new library makes it easier and more reliable to create themes for KWin, the Plasma window manager. It has impressive memory, performance, and sustainable development. If you miss the feature, don't worry. It will come back to Plasma 5.3.

KDE Plasma 5.3

April 28, 2015: KDE releases Plasma 5.3, adding a few new features and several bug fixes.

Improved Power Management

- Power management settings can be configured differently from other functions.

- The laptop cannot be paused when closing the lid while the external monitor is connected.

- Power control barriers prevent screen locking again.

- Screen light changes have now been animated on most computer systems

- It is not stopped when you close the lid while turning it off

- Support for keyboard button light controls on the lock screen

- KIinfoCenter provides statistics on power consumption

- The battery monitor now shows which apps are currently holding the power barrier.

Plasma Widgets

- The clipboard applet gains support by displaying barcodes

- The desktop codecs for desktop content and Folder View have been merged and seen performance improvements

- Recent documents and the latest application categories in the Application Menu (Kicker) are now enabled by KDE activities.

- Rewards of comic widget

- Return of plasmoid monitoring systems, such as CPU Load Monitor and Hard Disk usage

Plasma Media Center has been added as a preliminary technical test for this release. Fully stable but miss a few features compared to version 1. You can go directly to the Plasma Media Center session if you want to use it on a media device such as a TV or projector or use it from Plasma Desktop. It will scan videos, music, and photos on your computer to allow you to browse and play them.

KDE Plasma 5.4

August 25, 2015: KDE releases a feature to release a new version of Plasma 5.4. Its Plasma release brings users many good touches, like advanced DPI support, KRunner automation complete, and beautiful Breeze icons.

New Volume Applet

Our new Audio Volume applet works directly with PulseAudio, the popular Linux audio server, to give you complete control of the volume and output settings on a well-designed interface.

Another Dashboard Launcher

Plasma 5.4 introduces an entirely new full-featured Application Dashboard with KDE Plasma-addons: Includes all app menu features, including sophisticated measurement to screen size and full local keyboard navigation. The new launcher lets you easily and quickly find applications and recently used documents or favorites and contacts based on your previous work.

Artwork Galore

Plasma 5.4 delivers more than 1400 new icons that integrate all KDE applications and provide Breeze-themed technology to programs such as Inkscape, Firefox, and LibreOffice that offer a cohesive, native feel.

KDE Plasma 5.5

On December 8, 2015, KDE unveiled a new version of Plasma 5.5

Updated Breeze Plasma Theme

The Breeze Plasma widget theme has been updated to make it more consistent. While the Breeze icon theme adds new icons and updates the existing icon set to enhance the visual design.

Plasma Widget Explorer

The Plasma Widget Explorer now supports a two-column view with the new Breeze, Breeze Dark, and Oxygen widget icons.

Extended Feature Set to App Launcher

The context menus in the app launcher ("Kickoff") can now list recently opened documents in the app, allowing you to edit the app menu installation and add the app to the panel, Task Manager, or desktop. Favorites now support system texts, references, and system actions or can be created with search results. These (and other) features were previously only available in another Application Menu ("Kicker") and are now available in the Default Program Launcher for sharing backgrounds between the two launchers.

Color Picker

The color picker applet allows you to select a color from anywhere on the screen and automatically copy its color code to the clipboard in various formats (RGB, Hex, Qt QML rgba, LaTeX).

User Switch

User switches have been updated and upgraded and are now accessible in the app launcher, the new User Switcher applet, and the lock screen. It displays the full username and user set of the avatar.

KDE Plasma 5.7 Beta

June 16, 2016: DE releases a beta update to its desk-top software, Plasma 5.7. This release introduces a new entry-level screen design that completes the Breeze sensor we tried on Plasma 5.6. The structure has been redesigned and is more suitable for operating channels part of a domain or corporate network.

The Air and Oxygen Plasma themes that still support users opt for a three-dimensional design have also been upgraded. Plasma 5.7 calendar view on the calendar, providing a quick overview and easy access to upcoming appointments and holidays.

KDE Plasma 5.8

October 4. 2016: KDE releases a beta update for its desktop software, Plasma 5.8.

This release brings a new entry-level screen design that gives you a complete Breeze launch to almost indulge you. The structure has been redesigned and is more suitable for operating channels part of a domain or corporate network. Although very simple, it also allows for significant customization: for example, all Plasma wallpaper plugins, such as slideshows animated wallpapers, can now be used on the lock screen.

Support for Semitic written languages from right to left, such as Hebrew and Arabic, has dramatically improved. The content of the panels, desktop, and configuration dialogs is displayed in this setting. Plasma sidebar bars, such as widget checker, window switch, task manager, appear on the right side of the screen.

Visual desktop switch ("Page") and rewritten window list applet, using the new functionality of the task manager presented in Plasma 5.7. It allows them to use the same database as a task manager and improves their performance while reducing memory usage. The desktop switcher also could display the current screen only on multiple screens, and now we are sharing its multiple codes with the task switch applet.

KDE Plasma 5.8

October 10, 2017: KDE releases a beta update for its desktop software, Plasma 5.8. Plasma 5.11 delivers advanced notifications and a more powerful task manager, a redesigned settings app. Plasma 5.11 is the first release featuring a new "Vault." This system allows users to encrypt and open text sets in a secure and easy-to-use way.

New System Settings Design
A new design is added as an option; users who prefer an old icon or tree view can return to their favorite navigation trail.

Task Manager Development
New functions make it easier for applications to provide access to internal functions (such as text editor timetables, application modification options or document status, etc.), depending on what the application is currently doing. Additionally, rearranging windows in group pop-ups is now possible, allowing the user to order their open programs predicted. Despite all these changes, the function of the task manager has been improved to make it more efficient.

The new Plasma Vault offers solid encryption features introduced in an easy-to-use way. Plasma Vault allows you to lock and encrypt a set of documents and hide them from being seen even by a user. These "vaults" can be removed, encrypted, and opened quickly. Plasma Vault extends the feature of Plasma functions with secure storage.

INSTALLATION

Steps to Install KDE

The tasksel package will be used to install KDE Plasma in our Ubuntu programs.

- **Installation of Tasksel**
  - Tasksel is an Ubuntu package that provides a visual interface that allows users to install packages on their systems as if performing a specific task. In order to use the tasksel, we first need to install it on our systems. To do this, open the terminal by pressing the Ctrl + Alt + T keys or use the dash to gain access to a list of all installed applications. After opening the terminal, enter the following command:

    ```
    $ sudo apt install tasksel
    ```

  - To ensure that the tasksel is installed, enter the following command at the terminal:

    ```
    $ sudo tasksel
    ```

  - If you see a screen like the one shown below, it means that tasksel is installed on your system. Press Esc to return to the terminal.

- **Installing KDE Plasma**

  - Once the tasksel is installed, our next step is to install the KDE Plasma Desktop Environment in our Ubuntu system. There are two types of Plasma available for installation – small and full. The smaller version will end up with the Plasma Desktop environment. No other apps are installed, and users can install whatever they want later. This version is very useful if users do not want to use most of their memory or if users want to stick to the default Ubuntu apps.

  - To install this version, enter the following command at the terminal:

    ```
    $ sudo tasksel install kde-Plasma-desktop
    ```

  - The full version comes with a full KDE package, with all the main applications and desktop space of Plasma. For users who want to experience the full KDE, this version will be more suitable than their counterparts.

  - To install this version, enter the following command at the terminal:

    ```
    $ sudo tasksel install kubuntu-desktop
    ```

  - During installation, it will display information that asks you to edit sddm, which is KDE's display manager. After complete installation, you need to restart your system and a login screen will appear.

Login screen of KDE Neon.

- Enter your username and password to access your system, and a black screen will appear with the following icon.

- The first screen of the KDE desktop is as given below,

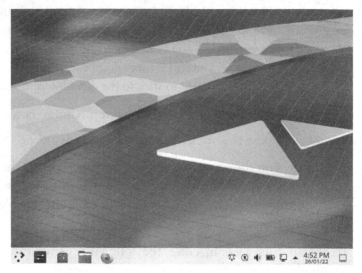

First screen of KDE.PNG.

APPLICATION

Application	Description
KSysGuard icon	System Monitor
K3b	Disk Burning
KolourPaint	Paint Program
KWrite	Text Editor
KColorChooser	Color Chooser
KFind	Find Files/Folders
KRuler	Screen Ruler
KCalc	cientific Calculator
Discover	Software Center
Plasma System Monitor	System Monitor
KBlocks	Falling Blocks Game
KNetWalk	Network Construction Game
Kigo	Go Board Game
Palapeli	Jigsaw puzzle game
KSquares	Connect the dots to create squares
Kubrick	3-D Game based on Rubik's Cube
KAtomic	Sokoban-like Logic Game
KBounce	Ball Bouncing Game

FEATURES OF KDE

KDE Plasma has recently been flooded with features, some of which even long-term Plasma users may not be aware of. Here is a list of KDE Plasma features that you may not know.

KRunner

It is a fantastic feature of Plasma and one of my favorites, but it is not known if it exists. There are ways to make it work but a quick keyboard shortcut. The default shortcut is Alt + Space, and once activated, you can do all sorts of things. You can launch apps, search for files, run commands, and more.

Quickly Move the Window

When you press and hold the Super key down, click on your mouse anywhere in the window and drag it to rotate the window very quickly. For this, you do not need to use the title bar or worry about accidentally clicking something because it only takes you to Cheat Window Mode.

Resize Window Quickly

If you press and hold the Super key down and right-click on your mouse near the edge of the window and drag it around, you can change the size of the window very quickly. Gone are the days of finding your mouse right on the edge of a window. As long as you are near the edge, this will work.

Zoom In and Out

Another nice feature of KDE Plasma is zooming in and out on your desktop. It is also an easy-to-use feature for those with visual impairments. Press and hold the Super button and use the Equals (=) button to zoom in on the image or the Minus (-) button to zoom out of your desktop.

Visible Desktops

Virtual Desktop is a fantastic feature common in the Linux ecosystem, although you may have heard of it called Workspaces elsewhere. By default, KDE Plasma comes with one Virtual Desktop. If you go to System Settings -> Workplace Conduct -> Visual Desktops, you can add as many as you want.

Desktop Grid Effect

Once you have a few visual desktops, you can look at the very cool effect of Desktop Grid. It does the same thing as Present Windows but instead does it for your visual desktops. You can open the Desktop Grid with the keyboard shortcut Ctrl + F8.

Small Desktop Menu

There is a feature that most people usually like in the Openbox window manager where you can right-click the desktop and have quick access to your Main Menu. To open this quick app menu, right-click your desktop and select "Fix desktop and background image." In the sidebar of the download window, select "Mouse Actions." Next, in the drop-down menu next to the Middle Button option, change that to "Application Launcher," You will now have a small desktop menu similar to Openbox.

CHAPTER SUMMARY

In this chapter, we have introduced KDE Plasma and its features along with history, core projects, applications, and development. We provided a separate section for beginners so they can develop an understanding of KDE version history.

CHAPTER 3

GNOME Desktop Environment

IN THIS CHAPTER

> Introduction

> Gnome

> GNOME-based linux distributions

> History of GNOME

> Pros and cons

We have already learned about Kool Desktop Environment (KDE) Plasma. In this chapter, we will briefly discuss one of the best flavors of Ubuntu named GNOME (GNU Network Object Model Environment). Primarily, it is an official flavor of Ubuntu and features the GNOME desktop environment. The Ubuntu flavor GNOME is a mostly pure GNOME desktop experience built from the Ubuntu repositories. Its first (unofficial) release was 12.10 (Quantal Quetzal), released in October 2012. The Ubuntu GNOME team announced their first official release Ubuntu Derivative: Ubuntu GNOME 13.04 in April 2013.

INTRODUCTION

To understand the concept of Ubuntu GNOME, let's begin with Ubuntu and then we will discuss the desktop environment GNOME.

DOI: 10.1201/9781003308676-3

We will cover basic terms before going deep into the GNOME desktop environments, including GNU/Linux, Open Source, Free Software, Graphical User Interface (GUI), Terminal User Interface (TUI), and CLI.

What Is Distribution?

The term "distribution" refers to the combination of kernel packages with the GNU libraries and applications. Ubuntu is one such distribution. It contains the Linux kernel, GNU tools, and many other applications and libraries.

Open-Source Linux Desktop Environment

The phrase "Open Source" is attributed to the Linux community which brought it into existence along with the introduction of Linux. "Linux" came into existence based on kernel. Many people and communities have contributed toward making Linux a complete Operating System which could replace UNIX.

Free Software

"Free software" is a software that respects users' freedom and community. It means that the users have freedom to do anything such as run, copy, distribute, study, change, or improve the software. Therefore, "free software" is a topic of liberty, not price. A free software program is on that adequately gives users all of those freedoms. Otherwise it is not free.

Key Points

- The freedom to run the program as per your wish
- Free software can be commercial
- The freedom to get the source code and make changes
- Legal considerations
- Contract-based licenses

Next, we will discuss the terms which are also related to the Ubuntu desktop environment KDE: GUI, CLI, and TUI.

The following section examines the GUI and the significant components of the Linux GUI. You will learn about standard window managers and desktop environments used with Linux.

Graphical User Interface

GNOME is the default GUI for most Ubuntu installations and is (loosely) based on the Apple ecosystem. A GUI, or graphical application, is anything you can interact with using your mouse, touchpad, or touch screen. There are various icons and other visual prompts that you can activate with a mouse pointer to access the functionalities. DE provides the GUI to interact with your system. You can use GUI applications such as GIMP, VLC, Firefox, LibreOffice, and file manager for various tasks.

Features of Linux GUI

The interface allows users to interact with the system visually with icons, windows, or graphics in a GUI. The kernel is the heart of Linux, whereas the GUI is the face of the operating system provided by the X Window System or X.

The product of the X.Org Foundation, an open-source organization, X Window System, is a protocol that allows links to be built on their X Server. You can use the X in one of the many window managers or desktop environments, such as the GNOME or the KDE. The desktop space includes a window manager and is a much more integrated system than a window manager. Built on a window manager, it requires both X Windows and a window manager.

Features of GUI

There are unique features and tools to interact with the software to make the GUI easy to use:

- **Button:** A graphical representation that acts as a program when pressed.

- **Dialog box:** A window type that displays additional information and asks the user for input.

- **Thumbnail:** A small representation of a program image, feature, or file.

- **Menu:** A list of commands or options provided by the user through the menu bar.

- **Menu bar:** A small, horizontal bar containing menu labels.

- **Ribbon:** Set up a file menu and toolbar that integrates program functions.

- **Tab:** A clickable area at the top of a window that shows another page or location.

- **Toolbar:** The Button Bar, usually near the app window's top, controls software operations.

- **Window:** A rectangular section of a computer display that shows the operating system.

The GUI uses icons, windows, and menus to execute commands, such as opening, deleting, and moving files. Although the GUI operating system is navigated using the mouse, the keyboard shortcuts or arrow keys can also be used.

For example, if you want to open an application on the GUI system, you can move the mouse pointer to the system icon and double-click it. With the command-line interface (CLI), you will need to know the commands to go to the program's directory, enter the list of files, and then use the file.

Benefits of GUI

A GUI is considered more user-friendly than a text-based CLI, such as MS-DOS, or the shell of operating systems like Unix. Unlike command-line or CUI operating systems, such as Unix or MS-DOS, GUI operating systems are easy to read and use because commands do not need to be memorized. Furthermore, users do not need to know programming languages. Thanks to its ease of use and modern appearance, GUI operating systems dominate today's market.

Command-Line Interface

CLI is a command-line program that accepts inputs to perform a particular function. Any application that you can use via commands in the terminal falls into this category. CLI is an old way of working with apps and applications and is used to perform specific tasks that users need. CLI is a text-based visual interface, unlike the GUI, which uses graphics options that allow the user to interact with the system and apps. CLI allows the user to perform tasks by entering commands. Its operating system is straightforward but not easy to use. Users enter a command, press "Enter," and wait for a response. After receiving the command, CLI correctly evaluates it and displays the output/effect on the same screen. The command-line interpreter is used for this purpose.

CLI is introduced with a telephone typewriter. This system was based on batch processing. Modern computers support CLI, batch processing, and a single-interface GUI. To make good use of CLI, the user must enter a set of commands (one by one) immediately. Many applications (mono-processing systems) still use CLI on their operators. In addition, programming languages like Forth, Python, and BASIC provide CLI. The command-line translator is used to use a text-based interface.

Another feature of CLI is the command line used as a sequence of characters used in the user interface (UI) or shell. Command information is used to inform users that CLI is ready to accept orders. MS-DOS is an example of CLI.

Terminal User Interface

TUI is also known as a Text-based User Interface. You have text on the screen because they are used only in the terminal. These applications are not well-known to many users, but there are a bunch of them. Terminal-based web browsers are an excellent example of TUI programs. Terminal-based games also fall into this category. Text User Interface (also known as written user interaction or terminal user interaction) is a text-based user. TUIs differ from command-line communication in that, like GUIs, they use all of the screen space and do not provide line-by-line output. However, TUIs use only the text and symbols found in the standard text terminal, while GUIs typically use high-definition image terminals.

GNOME

GNOME is an open-source movement, part of the GNU project and free software. It is similar to the Windows desktop system that works on UNIX and UNIX-like systems. It is not dependent on any other window manager. The current version runs on Linux, FreeBSD, IRIX, and Solaris.

The GNOME project provides two things:

1. The GNOME desktop environment, a built-in and attractive desktop for users.

2. The GNOME development platform, a comprehensive framework for building applications that integrates into the rest of the desktop.

So GNOME is a software that helps users and experts to develop a desktop and software that can be used for GNU.

Ubuntu GNOME Remix

Ubuntu GNOME is called GNOME Remix. It is a Linux distribution distributed as an open-source and free software. It is used as a GNOME 3 desktop environment instead of the Unity graphical shell with the GNOME shell. It became a "flavor" of the Ubuntu OS, starting with the 13.04 version and was announced in April 2007 as 17.04. Starting with its 17.10 version, the distribution was canceled in the standard Ubuntu favor, which changed from devoting Unity to GNOME Shell because of its desktop environment. But it is no longer supported.

What Is GNOME GNU?

In 1984, the GNU project was established to develop a complete Unix-like operating system with free software, i.e., the GNU system.

GNU kernel is not complete, so GNU is used with Linux. GNU and Linux are combined with the GNU/Linux operating system, now used by millions.

GNU stands for "GNU, Not Unix." It means that it is an independent system that uses Linux kernel and provides an open-source operating system to users.

What Is Ubuntu GNOME in Linux?

Linux is a free Unix-type operating system initially designed by Linus Torvalds with other developers around the world. It was developed under the GNU, which is a General Public License (GPL). The source code for the Linux distro is freely available to everyone over the Internet.

The last GNOME 3 version in its latest iteration is currently one of the most popular desktop environments used by almost every major Linux distro. It offers a modern desktop that delivers an intuitive user experience for all users – programmers and non-programmers alike.

The above lines indicate that it is an operating system used under GNU General Public License and provides various distros as open source for users. You can use these Linux distributions on the system as your primary operating system:

- Debian
- Fedora
- Manjaro

- openSUSE

- Solus

More about GNOME-Based Linux Distributions

GNOME3 is the default desktop environment on various major Linux distributions, including openSUSE, Fedora, Debian, Ubuntu, SUSE Linux Enterprise, Red Hat Enterprise Linux, CentOS, Pop!_OS, Oracle Linux, Endless OS, and Tails, as well as Solaris, a Unix operating system. The continued fork of the last GNOME 2 release, called MATE, defaults on many Linux distros that target low usage of system resources.

It is available for installing the Linux/GNU distributions. Several distributions provide the opportunity to try their demo before we install it. Some of them are explained below.

What Is openSUSE?

It is an origin of the original SUSE Linux distribution and a community-based distribution in contrast to SUSE Linux Enterprise. The SUSE company is still an influential sponsor of openSUSE. The relationship is similar to Fedora, CentOS, and Red Hat Enterprise Linux. The project uses a gecko logo to show the relationship between SUSE and openSUSE.

It is available in two flavors: the rolling-release Tumbleweed and the stable Leap. The latter is the same as Arch Linux as it is more of a "bleeding-edge" distribution with new software. You can install openSUSE as an old Linux system, but it is also available in the Windows Store for WSL. It may be neglected compared to other major Linux distributions, but it has a unique set of features and a code base with a wealthy estate.

openSUSE also gives you access to vanilla GNOME. But unlike Fedora, it follows a much slower release schedule. You won't get access to all the latest GNOME features as soon as they are released. However, this isn't technically a bad thing. It can dedicate more time and effort to make the OS more stable and reliable by having a slow release cycle. It makes it a perfect fit for professionals who can't afford to have their system crash in the middle of a meaningful work.

Now, openSUSE is distributed under two release models: Leap and Tumbleweed.

Each major version is released every three years, whereas point releases or minor updates are released annually with Leap. Depending on how often you want to upgrade your system, you should pick the flavor.

The openSUSE project offers two distributions.

1. Tumbleweed, which is a rolling distribution.

2. Leap, which is a point distribution.

What Is openSUSE Used For?

It is a project that promotes the benefit of free and open-source software. Its Linux distributions are well-known, mainly Tumbleweed, a tested rolling release, and Leap, a distribution with long-term support.

However, starting and switching applications work differently from other desktop operating systems. It only uses a single panel at the top of the screen. Its session is started on Wayland. openSUSE Leap uses GNOME with Wayland as default.

What Is Fedora?

Fedora gives GNOME, and it is an Open-source Operating System built and maintained by a community using the Linux kernel architecture. We can easily install it and use it live. Workstation 33 of Fedora is available now and ships GNOME 3.38 version I Fedora, the Only Linux Distribution, creates an operating system that is free to download, use, and modify as per your needs. All the features, software, packages, and components included inside are free. The Fedora community has thousands of volunteers, supporters, users, and contributors who interact via various online forums, email, and wikis to support each other. It provides the latest technology on recent hardware platforms with rapid development and release cycles.

The default desktop of Fedora is GNOME. Still, you prefer an alternative desktop environment such as KDE Plasma Desktop or Xfce. In that case, you can download a spin for your preferred desktop environment and use that to install Fedora, pre-configured for the desktop environment of your choice.

Fedora offers the latest GNOME experience out of all other Linux distros. It supports an ecosystem that provides users with new and updated software as soon as they are available.

With each new release of Fedora, the development team incorporates the latest version of GNOME. It allows you to access all the newly released GNOME features before anyone else.

However, since you get access to all the latest software first, there hasn't been much time to test them thoroughly. As such, be prepared to face the occasional bugs as new updates roll out. This is why Fedora is more suited for enthusiasts and computer geeks than regular users looking for stability.

What Is Debian?

Debian is the oldest Linux distro with tons of forks and derivatives. It officially supports almost all the major Linux distros, including GNOME, which is used by default.

Using Debian, you will get to enjoy GNOME in its purest glory. But keep in mind that you won't get updated to the newest releases as soon as they are available, as is the case with Fedora. But at the same time, releases aren't as slow as openSUSE or CentOS.

Overall, it is a highly stable and dependable Linux distro, perfect for beginners and advanced users. You will find Debian being seamlessly used by regular users to run day-to-day tasks and on web servers for hosting websites and web apps.

Also, Debian is the most massive community-run distro. When you combine that with the fact it has been around for so long, you get access to the best hardware and software compatibility.

What Is CentOS?

CentOS (Community Enterprise Operating System) is similar to open-SUSE as it focuses more on stability than delivering all the latest updates and releases. You won't get all the newest GNOME features, but you can rest assured that you rarely face bugs or system crashes.

The GNOME 3 desktop on CentOS 7 will provide a GUI for working with the Linux system. While I don't suggest using a GUI on a production server, it's a good option if you're using CentOS as a desktop.

As such, you will mostly find CentOS being used in enterprise situations. It is the most widely used platform in web hosting. It is also preferred by developers and large corporations looking for a mature and reliable OS with a longer release cycle.

What Is Arch Linux?

Arch doesn't ship with GNOME out of the box. It ships with no desktop environment at all. It is a super lightweight and flexible Linux distribution that allows you to build your own custom Linux experience with tools and software that you like. All you need to do is install the GNOME shell on your Arch system, and you are good to go. And since the OS doesn't come with any additional customization or extensions, you can rest assured that you will be getting the purest experience.

But that being said, you need to understand that you will need to install GNOME, and it won't be done for you. Furthermore, if anything goes wrong with the OS, you need to be knowledgeable enough to fix it yourself or find the solution through online forums.

Even installing Arch Linux can be far more intimidating than other Linux distros.

What Is Manjaro GNOME Edition?

Manjaro is based on Arch Linux and is available in many flavors, including a GNOME edition. The beauty of Manjaro is that you are getting access to the freedom and flexibility of Arch, but with GNOME already baked in. You don't need to worry about installing GNOME and other dependencies separately, making the process much more convenient and user-friendly.

With the Manjaro GNOME edition, you are getting an up-to-date GNOME shell desktop along with helpful software right out of the box. But that being said, the GNOME desktop on Manjaro is slightly customized, unlike with Arch, where you can get vanilla GNOME. Furthermore, it won't update you to the latest version of GNOME as soon as it's available, and you will need to wait a couple of weeks.

What Is Pop!_OS?

Pop!_OS is based on Ubuntu and built by System76 to be distributed along their computers. But now, it's a standalone product that you can download and install on any computer, not necessarily from their manufacturers.

The best thing about Pop!_OS is that it is ready to go as soon as you install it. For example, you get straight out-of-the-box support for AMD and Nvidia GPUs – you don't need to install any drivers manually. It makes it one of the best distros for gaming on Linux.

Like Ubuntu, it comes with a custom GNOME desktop, but it isn't as heavily skinned. On the contrary, Pop!_OS goes for a more minimal take, making GNOME feel even more sleek, intuitive, and beginner-friendly. This is why Pop!_OS is one of the most widely recommended distros for users who are just starting with Linux.

What Is Zorin OS?

It is another Ubuntu-based Linux distro designed with first-time Linux users in mind. It comes with a customized version of GNOME that allows you to switch between a Windows-like or a MacOS-like interface. Depending on which interface you pick, it will give you a familiar GUI that closely resembles the look and feel of your old OS to make the transition as seamless as possible.

You will also get access to Wine and PlayOnLinux out of the box. It allows you to run Windows applications on your Linux system, including Adobe software and the entire Microsoft Office Suite.

Thanks to Wine, all your favorite games that only have a Windows version will also run on Zorin OS, but it might not be as well optimized since it is running off a compatibility layer. It follows Ubuntu's long-term release cycle for security and stability, so you can expect a new Zorin OS version as soon as the next long-term Ubuntu release rolls out.

What Is Mageia?

We have Mageia, a fork of Mandriva Linux, currently defunct. KDE is the default desktop environment for Mageia, but you can pick GNOME as it is also officially supported, and it will give you a pure GNOME experience. It isn't as popular as the other distros on the list, but is gaining popularity – the release of its latest version Mageia 7.1 ships with tons of nifty bells and whistles. Mageia is super lightweight and easy to use for starters, making it highly beginner-friendly. But at the same time, it is also very flexible and offers tons of features that seasoned Linux users will appreciate. It supports a vast repository of software, including tons of productivity apps and games, so that you can run pretty much anything on the distro.

Furthermore, it is entirely processor agnostic, which is compatible with AMD, Intel, and even VIA processors. It is also very forgiving of your hardware configuration and will give you the best possible experience even if you are running it on limited specifications.

What Is Ubuntu?

Ubuntu 20.04 LTS version contains GNOME 3.36 with minor changes, and Ubuntu 20.10 has GNOME 3.38 version with minor changes. When we install a GNOME session, we can select for launching the new GNOME through the login screen.

If you are getting into Linux, you surely must have heard about Ubuntu. It is the most popular Linux distro. It is so popular that most non-users think it is synonymous with Linux.

Back in the day, Ubuntu came out with its custom desktop environment – Unity. But, as of Ubuntu 17.10, Canonical (the developers behind Ubuntu) has switched to the GNOME shell.

That being said, Ubuntu uses a heavily modified version of GNOME to maintain the design aesthetics of their Unity desktop. It might be a good thing for longtime Ubuntu users, but it isn't appreciated by users looking to get the GNOME experience as its developers intended.

System requirements of Ubuntu GNOME:

- The memory of the system should be 1.5 GB of RAM.

- The system should have a 1 GHz processor (like Intel Celeron) or better.

- Access to the Internet is helpful (to install updates during installation).

- You can use a USB port or DVD/CD drive for the installer media.

Various distros specialize in different fronts, so you can pick one that resounds to your needs and requirements. If you are looking for vanilla GNOME, go with Fedora or Arch, both have access to all the latest features as soon as they are released. On the other hand, if you are looking for a little more stability, then Debian, openSUSE, and Mageia are perfect alternatives, with CentOS being the most stable and reliable with a long-term release cycle.

However, assume you want to stay in the middle and access new features in a reasonably timely fashion without sacrificing stability. In that case, you can test out Manjaro or POP!_OS, both of which are incredibly beginner-friendly. And finally, if you want to use GNOME because of its features and are not concerned about how it looks, both will provide you with a heavily customized GNOME desktop but are filled with valuable features and welcome new users.

HISTORY OF GNOME

The project started as the unofficial "remix" because a few users prioritized the GNOME 3 desktop on Unity. The 12.10 version of GNOME Quantal Quetzal was the initial version published on October 18, 2012. The founder of Ubuntu, Mark Shuttleworth, and Canonical Executive Chairman declared on April 5, 2017, that the Ubuntu mainline version would move through Unity to the GNOME 3 desktop starting with 18.04

LTS version. It makes it identical to Ubuntu GNOME 3. After that, it was revealed that the 17.10 version of Ubuntu would be the first version for using GNOME.

In April 2017, Mark Shuttleworth specified that "Ubuntu GNOME along with the intent of distributing a fantastic each GNOME desktop and the team of Ubuntu GNOME support, not making it competitive or different with that effort." After the announcement of Ubuntu that they will switch the desktop environments via Unity to GNOME, the GNOME developers revealed on April 13, 2017, that distribution will merge within the Ubuntu, which is starting with the 17.10 release.

GNOME 3 is now the default desktop environment on many Linux distributions, including Ubuntu, SUSE Linux, Red Hat Linux, Fedora, Debian, CentOS, Oracle Linux, Endless OS, Tail, and Solaris, a Unix operating system. The fork of the last GNOME 2 release is called MATE.

The GNOME 1 release looked very similar to Windows 98, a wise decision that immediately provided a familiar graphical interface for new Linux users. GNOME 1 offered desktop management and integration, not only simply window management. The files and folders can drop on the desktop, providing easy access. It was a robust advancement, and many major Linux distributions included GNOME as the default desktop. Finally, Linux had an actual desktop. We will discuss GNOME later in the other section.

Over time, GNOME continued to evolve. In 2002, GNOME 2 was a significant release. They cleaned up the user interface and tweaked the overall design. Instead of a single toolbar or panel at the bottom of the screen, GNOME 2 used two panels at the top and one at the bottom. The top panel included the GNOME Applications menu, an Actions menu, and shortcuts to frequently used applications. The bottom panel provided icons of running programs and a representation of the other workspaces available on the system. Using the two panels provided a cleaner user interface, separating "things you can do" (top panel) and "things you are doing" (bottom panel).

Many other users felt the same, and GNOME 2 became a standard for the Linux desktop. The successive versions made incremental improvements to GNOME's user interface, but the general design concept of "things you can do" and "things you are doing" remained the same.

Despite the success and broad appeal of GNOME, the GNOME team realized that GNOME 2 had become difficult for many to use. The application's launch menu required too many clicks. Workspaces were challenging to use. Open windows were easy to lose under piles of other application windows. In 2008, the GNOME team embarked on updating the GNOME interface. That effort produced GNOME 3.

GNOME 3 removed the traditional taskbar in favor of an Overview mode that shows all running applications. Instead of using a launch menu, users start applications with an Activities hot button in the black bar at the top. Selecting the Activities menu brings up the Overview mode, showing both the things you can do with the favorite applications launcher to the left of the screen and the things you are doing with window representations of open applications.

Since its release, the GNOME 3 team has improved it and made it easier to use. GNOME 3 is modern, familiar, striking, and can balance features and utility.

GNOME 1 (1999)

GNOME was launched by Miguel de Icaza and Federico Mena on August 15, 1997, as a free software project; they developed a desktop environment and applications for it. It was founded partly because K Desktop Environment, growing in popularity, relied on the Qt widget toolkit, which was used as proprietary software license until version 2.0 (June 1999). In place of Qt, GTK (GNOME Toolkit, called GIMP Toolkit) is the base of GNOME. GTK uses the GNU General Public License, a free software license that allows software linked to it to use a much more comprehensive set of permissions, including proprietary software licenses. GNOME is licensed under the LGPL for its libraries and the GNU General Public License for its applications.

"GNOME is a flexible Graphical User's Interface that combines ease of use, the flexibility, reliability of GNU/Linux. We are extremely excited about GNOME and mean for the future of GNU/Linux computing," Miguel de Icaza, chief designer of GNOME said.

GNOME is designed to be portable to the modern UNIX system. It runs on Linux systems, BSD variants, Solaris, HP-UX, and Digital Unix. It will be included in Red Hat and other Linux distributions such as Debian GNU/Linux and SUSE Linux.

It has features that allow users to assign an icon to a file or URL. It has a drag and drop–enabled desktop, using the standard Xdnd and Motif protocols.

Its code makes it easy for international users, with core components recently supporting more than 17 languages, with more on the way. It works well with various scripting and compiling languages, including Ada, C, C++, Objective-C, TOM, Perl, and Guile.

"GNOME is a big step towards acquiring the Free Software Foundation's providing a whole spectrum of software from experts to end-users. We all are excited about the direction of GNOME taking us in," Richard Stallman, founder and president of the Foundation, said. "'Free Software' includes the freedom to run, copy, distribute, study, change and improve any software distributed under the General Public License. It will create an energetic environment for programmers and users to create use GNU/Linux and GNOME programs. We will see a wide range of GNOME-based applications to answer the different needs of computer users."

GNOME is available for free at http://www.gnu.org, http://www.gnome .org, and several other mirror sites. It is also included in Red Hat Software, Inc.'s GNU/Linux distributions, with other sources available in the coming years.

GNOME 2

It was released in June 2002 and was very similar to a traditional desktop interface, featuring a simple desktop in which users could interact with virtual objects, such as windows, icons, and files. It was started with Sawfish as its default window manager but later switched to Metacity. In GNOME 2, the concept of handling windows, applications, and files is similar to that of modern desktop operating systems. In the default configuration of GNOME 2, the desktop has a launcher menu for rapid access to installed program file locations; a taskbar can access open windows at the bottom of the screen. The top-right corner features a notification section for programs to display notices during running in the background. Hence, all these features can be moved to any position or orientation the user wants, replaced with other functions, or removed altogether. GNOME 2 provides similar features as the conventional desktop interface.

GNOME 2 was the default desktop for OpenSolaris, and the MATE desktop environment is a fork of the GNOME 2 codebase.

GNOME 2.0.2 is the predecessor version of 1.0, which is a complete, accessible, and easy-to-use desktop environment. In addition to basic desktop functionality, it is a robust application framework for software developers, support for object embedding, and accessibility. GNOME 2 is part of the GNU Project and is a free software.

The GNOME 2.0.2 platform includes a complete library suite to support GNOME applications. It also includes all the essential utilities for your day-to-day computing, from a simple weather monitor to a powerful file manager. GNOME 2.0.2 is compatible with several platforms, including GNU/Linux, Solaris, HP-UX, Unix, and BSD.

The GNOME 2 development cycle allowed several features that improved performance and usability. It also includes a robust new framework that developers can leverage.

Some features of GNOME 2 are:

- Improved fonts and graphics

- Usability

- Performance

- Keyboard navigation

- Accessibility

- Internationalization

- XML

Improved Font and Graphics

- Fonts can be anti-aliased (familiar graphics setting).

- There is no flicker in GTK apps.

- Images are composited onto backgrounds with the entire alpha channel, accelerated via MMX and the RENDER extension.

- There are new enhanced icons.

Usability
Streamlining, consistency, and coherence are having the primary focus of GNOME 2 Usability work.

- **Streamlining**: GNOME 2 has been simplified, and there is no need to add a pipeline. The interface clutter controlled a GNOME where you could almost literally "do less with more." GNOME 2 removes many obscure or rarely used features. In exchange, you will find that most of the features you care about are much easier to access, because a million other items do not obscure them.

- **Consistency**: Interfaces that behave according to uniform patterns are easier to learn, faster to use, and less tending to error. The GNOME Interface Guidelines have helped make the GNOME 2 interface more predictable, producing consistency between applications and promoting usable patterns within particular applications.

- **Coherence**: The GNOME 2 desktop fits nicely from "Login" to "Log Out"; countless brainstorming hours and tireless hacking have produced a desktop and more than a loose confederation of modules.

Specific User-Visible Improvements Include

1. Menus and panel
2. Dialogs
3. Icons and themes
4. Applications

Menus and Panel

- The windows can be dragged between workspaces with the Workspace Switcher applet.

- The menus can scroll when they get too big.

- More innovative menus accommodate diagonal mouse movements.

Dialogs

- The file selector doesn't forget names when selecting a different folder.

- The updated color and font selectors.

- New dialog Run Program with command completion.

- The text fields include right-clicking menus for cutting, copying, and pasting text.

Icons and Themes

- The new stock icons and color palette.

- It supported theming of stock icons.

- The CD Player and login screens are now themeable.

- An attractive default appearance.

Applications

- Everything is redesigned and easier to use Search Tool.

- The brand-new lightweight help application, Yelp.

- It can control center applications for managing GNOME 2 properties that have been greatly simplified and reduced in number.

- It has increasing compliance with freedesktop.org standards.

- Rewritten terminal application with tabs and profiles.

- A brand-new dynamic, centralized configuration system.

- Many applications have been renamed to suit their purposes better.

GNOME 3

It was released in 2011. While GNOME 1 and 2 interfaces followed the classic desktop analogy, the GNOME Shell assumed a more abstract metaphor with streamlined window management workflow unified header bar that replaces menu bar, taskbar, and toolbar and minimize and maximize buttons hidden by default.

GNOME 3 brought many enhancements to the core software. Many GNOME Core Applications went through redesigns to provide a more consistent user experience, and Mutter replaced Metacity as the default window manager. Adwaita substituted Clearlooks as the default theme.

Features of GNOME

GNOME has become one of the most efficient, stable, and reliable desktops available for the Linux operating system. Not only that, it remains incredibly user-friendly. Most users have experienced this regardlessly and can get incredible speed with GNOME without applying any extra effort. So, let's look at the details of GNOME and know about its great features.

Let's Take a Look at Its Features

- **Flatpak**: It seems like one of the best things ever happened to Linux. It lets developers distribute an app on every Linux distribution with ease. The new GNOME includes GTK+ theme handling and language configuration support.

- **New Boxes Features**: Boxes built-in applications can run remote and locally installed virtual machines. So, you don't need to install any virtual machine such as VMware or VirtualBox to try other Linux distros. It can automatically download operating systems from the new box assistant; all you have to do is pick the Linux distro you want to use, and Boxes will do the rest.

- **Activities Overview**: One of the essential pieces of the GNOME puzzle is the Activities overview. Activities are where you access application launchers, minimize applications, search, and virtual desktops.

- **Multimedia Apps**: Photos App has a new import feature that allows you to easily add photos to your library from SD cards and USB drives. The app can now auto-detect storage devices with new images, giving you an option to organize the pictures into albums during import itself. Other multimedia enhancements include playing MJPEG video files through a video player and reordering playlists by drag and drop in the music app. Games app has an exciting feature. It has an excellent new CRT video filter that makes game visuals look like they are being played on an old CRT TV.

- **Dash**: If you don't want application launchers on the Dash, right-click the launcher in question and select Remove from Favorites. If you like to add a favorite to the Dash, you need to open the Applications overview. To do so, click on the grid icon at the bottom of the Dash. When the Applications open, you can scroll throughout the list of installed applications to find what you are looking for. You can run that application or add it to the Dash with a single click. To add an application to the Dash, right-click the application icon, and select Add to Favorites.

- **File Manager**: Numerous improvements in Files application. This feature allows batch renaming of files. Apart from that, compressed file functionality has also been integrated into Files. Many other user interface improvements have also been added.

- **Favorites in Files**: In Files, you can select files and folders and add them to the favorites list so that you can quickly view them in a "Starred" list, as shown above. The favorite capability has also been added to the Contacts application and can pin your favorite contacts with whom you interact more.

- **Night Light**: It is the main feature in GNOME. It works as advertised, subtly adjusting the color temperature of your monitor based on the time of day. During the day, you will see things as you used to. The screen temperature gets colder and brighter with more blue light – the screen transitions to a warmer hue with less blue light in the evening. The blue light filters help promote natural sleep cycles and reduce eye strain.

- **Calendar**: The month view in the Calendar in GNOME shows the Events more readable. You can also expand cells that have several events overlapped. Also, notice the weather info besides the events. The To-Do list has been revamped to reorder tasks by drag and drop.

- **Terminal**: Terminal gets some enhancements as well. Notice the redesigned preferences window. There are no longer separate Preferences; both Profile and Preferences options are clubbed into one window. You will also see blinking text.

- **Search**: The GNOME Search tool is potent. Not only can it search for installed applications within the Application overview, but it can also search for applications not yet installed within GNOME Software and search for files.

- **GNOME Beautification**: GNOME looks more sleek and beautiful with Cantarell's enhanced default interface font. Character forms and spacing have been improved to make the text more attractive.

- **A More Wonderful Web**: The Web is the default GNOME browser and is often overlooked by users preferring to browse the Web with a better-known app, like Mozilla Firefox or Google Chrome. Bookmarking web pages on the Web now takes a single click. A new bookmarks popover makes it easy to access existing bookmarks, and a new interface for managing, editing, and tagging bookmarks will appease those who like to stay organized.

- **Better Device Support**: GNOME comes with integrated Thunderbolt 3 connection support. Security checks are added to prevent data theft through unauthorized Thunderbolt 3 connections. The top bar shows the Thunderbolt 3 connection status when active. Touchpad uses a gesture for the secondary click, which is nothing but right-click action in a mouse. Keep a finger in contact with the touchpad and tap with another finger to use the motion.

- **Clocks App**: Now, adding the UTC time zone to your world times is possible.

- **Better Icons**: There's one area where the GNOME desktop often shorts. A slate of high-resolution icons is included to ensure that everything looks sharp and detailed on high-density displays. Many redesigned devices, mime-type, and app icons feature a brighter, cleaner, and more modern look.

- **App "Usage"**: GNOME ships with a new technology preview app called Usage. Using this App, you can see CPU and RAM consumption and highlighted problem areas. It is a great feature that helps you troubleshoot issues quicker than before. More features are planned in future updates to the App.

- **Screenshots**: Another great feature is built-in screenshot support. Driven by hotkeys, it allows you to capture the whole screen, an active window, or a screen region to the clipboard or directly to a PNG file under $HOME\Pictures\Screenshot-*.

- **All-New On-Screen Keyboard**: The on-screen keyboard has been recoded entirely in GNOME. The new keyboard is user-friendly and automatically pops up when a text is selected. The view gets away to ensure you can see what you are typing.

- **Minimalist Design**: Most interfaces include quite a few on-screen elements unrelated to the task at hand. Windows and Chromebooks have taskbars across the bottom that contain all the favorite or open apps. On GNOME, the panel at the top does not contain any app launchers. The panel is small and black, like on a phone or tablet. It includes the date and time and a few system indicators in the top right.

GNOME SHELL

It is the graphical shell of the GNOME desktop environment starting with version GNOME 3, released on April 6, 2011. It provides essential functions like launching applications, switching between windows, and a widget engine. GNOME Shell replaced GNOME Panel and some ancillary components of GNOME 2. GNOME Shell is written in C and JavaScript as a plugin for Mutter.

In contrast to the KDE Plasma Workspaces, a software framework intended to facilitate the creation of multiple graphical shells for different devices, the GNOME Shell is designed to be used on desktop computers with large screens operated via keyboard and mouse, as well as portable computers with smaller screens operated via their keyboard, touchpad, or touch screen. However, a fork of the GNOME Shell, known as Phish, was created in 2018 for specialization with touch screen smartphones.

History of GNOME Shell

The first version of GNOME Shell was created during GNOME's User Experience Hackfest 2008 in Boston. After the traditional GNOME desktop and accusations of stagnation and low concept, the resulting meeting led to GNOME 3.0 in April 2009 since Red Hat has been the main driver of GNOME Shell's development.

Pre-release versions of GNOME Shell were first made available in August 2009 and became a regular, non-default part of GNOME in version 2.28 in September 2009. It was finally shipped as GNOME's default user interface on April 6, 2011.

Software Architecture

GNOME Shell is integrated with Mutter, a compositing window manager, and Wayland compositor. It is based on Clutter to provide visual effects and hardware acceleration. According to GNOME Shell maintainer Owen Taylor, it is set up as a Mutter plugin primarily written in JavaScript and uses GUI widgets provided by GTK+ version 3.

Features

- The changes to the user interface include but are not limited to:

- Support clutter and Mutter multi-touch gestures.

- Support for HiDPI monitors.

- A new Activities overview, which houses:

  a. A dock or Dash is used for quickly switching between and launching applications

b. A window selector, similar to macOS's Mission Control, also incorporates a workspace switcher/manager

c. An application picker

d. Search

- "Snapping" windows to screen borders makes them fill up half of the screen or the whole screen.

- By default, a single-window button, Close, instead of three.

- The minimization option has been removed due to the lack of a panel to minimize. Maximization can be accomplished using the window above snapping or double-clicking the window title bar.

It provides core interface functions like launching applications, switching windows, or seeing notifications. It has the advantage of the capabilities of modern graphics hardware and introduces innovative user interface concepts to provide a delightful and easy-to-use experience. GNOME Shell is the technology of the GNOME 3 user experience.

A fallback mode is offered in versions 3.0–3.6 for those without hardware acceleration which offers the GNOME Panel desktop. The method can also be toggled through the System Settings menu. GNOME 3.8 removed the fallback mode and replaced it with GNOME Shell extensions that offer a more traditional look and feel.

GNOME Shell has the following graphical and functional elements:

- Top bar
- System status area
- Activities overview
- Dash
- Window picker
- Application picker
- Search
- Notifications and messaging tray
- Application switcher
- Indicators tray

Extensibility

The functionality can be changed using extensions, which is written in JavaScript. You can find and install extensions using the GNOME extensions website. Some of these extensions are hosted on the GNOME git, or are illegal.

INSTALLATION

The "Vanilla Gnome" version is a GNOME desktop installation.

1. To install the GNOME desktop environment, open a command line and use system package manager to install the GUI and gdm3 window manager.

```
$ sudo apt update
$ sudo apt install gnome-session gdm3
```

2. If the installation prompt asks you to select your default window manager (because you are currently using a different one), be sure to select gdm3 if you want GNOME to be your default desktop environment.

3. After installation is complete, reboot your system, and you will be presented with a GNOME login.

```
$ reboot
```

4. At this point, the GUI should start. You need to select desired desktop flavor on the login page before you login.

Or

The other way is to install full GNOME by using the tasksel command. First ensure that the tasksel is installed on your system:

```
$ sudo apt update
$ sudo apt install tasksel
```

Now, use the tasksel command to install GNOME desktop:

```
$ sudo tasksel install ubuntu-desktop
```

Reboot your Ubuntu system to be presented with the GNOME desktop:

```
$ reboot
```

The first window of GNOME is given below:

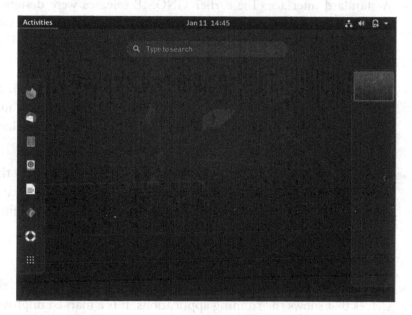

Desktop environment GNOME interface.

GNOME 3: Pros and Cons

Switching to GNOME 3 is both an opportunity and a distraction. On the one hand, it is the opportunity to put aside some annoying behaviors in earlier GNOME releases. On the other hand, GNOME 3 is a distraction because its changes can get in the way of long-established work methods.

As a result, you need to look at GNOME 3's pros and cons before deciding to make the new desktop part of your everyday computing unless, of course, you are the sort that automatically rejects or embraces change simply because it is unique.

GNOME 3 contains many changes. For example, you might see that improved hardware interaction that GNOME 3 offers a Suspend option only on a machine that supports that option. Such enhancements are easy to overlook and, despite their convenience, too minor to be a large part of anybody's reaction to GNOME 3.

Here are various pros and cons of the new desktop that might be important to you:

Pros

- A standard interface: The earlier GNOME releases were designed with the workstation and the laptop in mind. That is no longer realistic in this age of netbooks, tablets, and mobile devices. GNOME 3 is designed for all users.

- In the GNOME 2 series, the system settings menu of alphabetized items is divided into Personal and Administration sub-menus. GNOME 3 reduces inefficiency with a window of settings organized by category that is easier and quicker to scan.

- GNOME 3 replaces the menu with a list of applications on the Activities overview screen. These changes allow larger icons and eliminate the problem of editing to keep it short at the risk of effectively hiding items from users.

- Improved display of virtual workspaces.

- On the right of the Activities is a visual display of all open workspaces that shows the running applications. It is a marked improvement over earlier GNOME releases.

- The most significant advantage of GNOME 3 is most likely to be resisted. Hence, GNOME 3 makes several advanced features more prominent and easier to use.

- The Dash on the Activities displays your favorites more prominently, while switching between screens encourages learning keyboard shortcuts.

- GNOME 3 allows you to move to a messaging window without switching the focus.

Cons

- GNOME 3 doesn't allow icons on the desktop. It depends on the distribution. That will be irrelevant for half of the users, but this limitation will be a deal-breaker for the other half. You can learn to edit Gconf – and, so far, I haven't found any instructions on the Web

– they will either have to learn to live without icons or hunt for a new distribution.

- You can switch to the Activities screen to open applications. Selecting an application that immediately changes you to the workspace means that you have to switch back to the Activities page to open any application you want to run simultaneously. The limitation also exists in the classic menu of the GNOME 2 series, but it requires far more mouse clicks in GNOME 3.

CHAPTER SUMMARY

In this chapter, we covered GNOME and introduced some of its features. We discussed its history, core projects, applications, and development. We also provided a separate section on the history of GNOME's versions.

xfce Desktop Environment

IN THIS CHAPTER

> Introduction

> Versions history

> The xfce core desktop components

> xfce installation

> Advantages and disadvantages

After studying GNOME in the previous chapter, we will in this chapter briefly discuss the Linux-based operating systems named xfce. Primarily, it is an official Linux OS and has various features.

INTRODUCTION

Olivier Fourdan developed the xfce desktop environment, which began in late 1996. He began his career in technology production, web development, and embedded Linux systems. Fourdan has been working for Red Hat since 2007, interrupted for two years at Intel between 2013 and 2014. Since 2017, he has been active in adopting Wayland, working on many different components, including GTK, Mutter, GNOME Control Center, XWayland, and Mesa3D.

DOI: 10.1201/9781003308676-4

VERSIONS HISTORY

Earlier Versions

xfce started as a simple project made with XForms. Olivier Fourdan released the program, a simple taskbar, on SunSITE. Fourdan continued to develop the project, and in 1998, xfce 2 was released with the first version of xfce window manager, Xfwm. He requested that the project be included in Red Hat Linux but was rejected due to its XForms base. Red Hat only accepts open source software and is licensed under the GPL or BSD license, while, at the time, XForms was a closed and free source for personal use. For the same reason, xfce was not available to Debian before version 3, and xfce 2 was only distributed in the Debian storage area.

In March 1999, Fourdan began a complete rewriting of the project based on GTK, a non-patented tool kit that grew in popularity. The result was xfce 3.0, licensed under the GPL. The project received GTK drag and dropped support, native language support, improved customization, and full-featured free software. xfce was uploaded to SourceForge.net in February 2001, starting with version 3.8.1.

Modern xfce

xfce 4.4 desktop showing various Xfwm effects: drop shadows behind windows, alpha-sided windows, and panel.

In version 4.0.0, released on September 25, 2003, xfce was upgraded to use GTK 2 libraries. It has Built-in transparency and shadow cast and a default SVG icon set. In January 2007, xfce 4.4.0 was released. It includes Thunar file manager instead of Xffm. Desktop thumbnail support was added – also, various improvements were made to the panel to prevent buggy plugins from crashing the entire panel. In February 2009, xfce version 4.6.0 was released. This version has a new configuration backend, a new settings manager, and a new audio connector, and several important improvements to the session manager and other key components of xfce.

xfce 4.8.0 was released in January 2011. This version includes the ThunarVFS and HAL exchanges with GIO, udev, ConsoleKit, and PolicyKit, and new remote network browsing services using a few protocols, including SFTP SMB and FTP. The windows volume has been reduced by combining all Thunar file dialog boxes into one conversation. The panel application has been rewritten for better configuration, brightness, object management, and launcher 4.8, and introduces a new menu plugin to view the directory. The 4.8 plugin framework is always compatible with 4.6 plugins. The 4.8 display configuration dialog supports RandR

1.2. It automatically detects screens and allows users to select their preferred display setting, refresh rate, and display rotation. Many displays can work in clone mode or place next to one. Keyboard selection has been updated to make it easier and easier to use. Also, the manual settings are manually edited to make it more efficient.

The development cycle 4.8 was the first to implement a new release strategy developed after the "xfce Release and Development Model," developed at the Ubuntu Desktop in May 2009. A new web system was used to simplify release management with a dedicated Transifex server and edited by xfce translators. The project server and port infrastructure were also upgraded to meet the expected demand following the release of 4.8.

xfce 4.10 was released on April 28, 2012, that introduced the panel's direct display mode and distributed most of the wiki to the Internet. The main focus of this release was to improve the user experience.

xfce 4.12 desktop operating system on Fedora 22: note that the file manager has been rewritten in GTK 3. xfce 4.12 was released two years and ten months later, on February 28, 2015, contrary to online speculation about the project "dead." The goal of 4.12 was to improve user experience and make the most of the technology currently available. New window manager features include the Alt + Tab dialog box and intelligent capture for multiple monitoring. Also, a new panel notification panel plugin has been introduced, as well as a rewritten text editor and an advanced file manager. xfce 4.12 also upgraded to GTK 3 by deploying apps and supporting plugins and bookmarks. At 4.12, the project reaffirmed its commitment to platforms such as Unix other than Linux by installing OpenBSD screenshots.

xfce 4.13 is an upgrade release during the transition of the transport components to fully compliant with GTK3, including xfce-panel and xfce-settings.

The scheduled release of xfce 4.14 was announced in April 2016 and was officially released on August 12, 2019. The main release targets included transferring the remaining key components from GTK 2 to GTK 3; instead of relying on Dbus-glib via GDBus, GNOME implementation of DBus specification and extracting deleted widgets. Significant features have been postponed for later release of 4.16. The smaller version of GTK 3 was hit from 3.14 to 3.22.

xfce 4.16 was released on December 22, 2020. Some notable changes to this release include new icons with a color palette; advanced links to change system settings; various panel enhancements such as hide animation, a new notification plugin supported by both SysTray asset, and modern StatusNotifier features; and better support for black themes.

THE XFCE CORE DESKTOP COMPONENTS

1. exo

2. gtk-xfce-engine

3. libxfce4ui

4. libxfce4util

5. thunar

6. thunar-volman

7. xfce4-appfinder

8. xfce4-panel

9. xfce4-session

10. xfce4-settings

11. xfconf

12. xfdesktop

13. xfwm4

14. garcon

15. container

16. xfce4-power-manager

All critical components of the xfce desktop must comply with the removal policy described in this document.

USAGE

Display Managers

xfce4-session includes a file that should add an option to display managers to run an xfce session. xfce Desktop Environment does not have its DM, but various options such as gdm, slim, lxdm, and lightdm. Check out this link for details.

Command-Line

It uses startxfce4 to start an xfce session or select an xfce Session in the login manager, including a session manager, panel, window manager, and desktop manager. See automatic login from the console for more information. By default, the xfce session manager controls the implementation of applications. It allows saving your session when you quit xfce so that the next time you sign in, the same apps will be started automatically.

Features

xfce contains several key components of small tasks you can expect on the desktop:

- **Window Manager:** Controls the placement of windows on the screen, provides window decorations, and manages visual effects or desktops.

- **Desktop Manager:** Sets the background image and provides the root window menu, desktop icons or minimized icons, and a list of windows.

- **Panel:** Switches between open windows, open apps, switch apps, and menu plugins to browse apps or directories.

- **Time Manager:** Controls desktop login and desktop management and allows you to save multiple login times.

- **Application Finder:** Displays applications installed in your system in stages, so you can quickly find and launch them.

- **File Manager:** Provides basic file management features and various services such as bulk renaming.

- **Settings Manager:** Tools for controlling various desktop settings include keyboard shortcuts, appearance, display settings, etc.

In addition to the basic set of modules, xfce also offers many additional apps and plugins so you can customize your desktop the way you like, for example, terminal emulator, text editor, audio connector, app finder, image viewer, iCal-based calendar, and CD, and a DVD burning application. You can read more about xfce modules on the projects page. xfce project contains several different projects on each part of the desktop. Some projects have their project pages to provide additional information.

MAIN COMPONENTS OF XFCE

xfwm4 – Window Manager

In xfce 4, Window Manager is part of the xfce Desktop Environment. The command to run it in front is xfwm4; to launch it in the background, use the xfwm4-daemon. It is responsible for arranging windows on the screen, providing window decorations, and permitting them to be moved, resized, or closed.

Xfwm4 adheres strictly to the standards set on freedesktop.org. As a result, special features such as opening windows or providing an app icon should now be used in the application; you can no longer use the window manager to force different behavior. One of the best features of xfwm4 is easy handling; themes are available at xfce-Look. Window decorations (borders, title bar, and window buttons) can be adjusted using window manager themes. Xfwm4 provides multi-header support for both xinerama and multi-screen mode, helpful if several monitors are connected to your system. Xfwm4 can be run independently, but if used in this way, xfce Settings Manager will be required as a GUI to make configuration changes. Tasks other than managing windows, such as setting a background image or launching programs, need to be done with other modules.

Xfwm4 integrates with its integration manager, using the new X.org server extensions. Composer is like WM alone; it holds a wide variety of windows, monitors all types of X events, and responds appropriately. An integrated controller embedded in the window helps keep various viewing effects synchronized with window events. If you want to use a compiler, you have to create xfwm4 using the configuration-enable-compositor option. In either case, you can disable the composer at xfwm4 launch using the "-compositor = off" argument. It controls the placement of program windows on the screen, provides window decorations, and manages virtual workplaces or desktops.

Xfwm4 Behavior

- Give focus to a window
- Maximize/unmaximize windows
- Resize windows
- Hide/unhide windows

- Shade/unshade windows

- Stick/unstick windows

- Raise/lower windows

- Move windows

- Move a window to another workspace

- Close a window

xfce Panel – Panel

The panel is a bar that always allows program launchers, panel menus, clocks, desktop switches, and more.

It is part of the xfce Desktop Environment and includes application launchers, panel menus, workspace switches, and more. Many panel features can be customized with the GUI and GTK + style layouts and Xfconf hidden settings.

The panel will usually start automatically as part of your xfce session when you start the xfce Desktop Environment. To start the panel manually, you can use the xfce4 panel in the terminal. If the panel gets started for the first, it will give you the following options:

Use the Default Setting

This will introduce the default panel configuration with most automated plugins. It gives a good start to planning and making your configurations.

One Empty Panel

It will give you a blank single panel window, which is helpful once you know what kind of configuration you want.

Internal Plugins

There are various internal plugins used in the panel as given below.

- **Action Buttons:** Add system action buttons to the panel

- **Program Menu:** Add a menu containing installed program categories

- **Clock:** Panel clock plugin

- **Directory Menu:** Display a menu tree in the menu

- **Launcher:** An app launcher with a menu of your choice
- **Notification Location:** The location where notification icons appear
- **Separator:** It adds a separator or space between panel plugins
- **Show Desktop:** It can hide all windows and show desktop
- **Window Buttons:** It switches between open windows using buttons
- **Window Menu:** It switches between opening windows using the menu
- **Workspace Switch:** It switches between virtual desktops

External Plugins

There are various external plugins used in the panel as given below.

- **Battery:** xfce4 battery monitoring plugin compatible with APM and ACPI, Linux, and * BSD
- **Calculator:** xfce4 panel calculator plugin
- **Clipman:** xfce clipboard manager
- **Cpufreq:** Displays information about the CPU controller and the frequencies your system supports and uses
- **Cpugraph:** Provides multiple display modes (LED, gradient, fire, etc.) to display the current CPU load of the system
- **Date Time:** Shows the date and time on the panel, and a calendar appears when you left-click on it
- **Diskperf:** Displays disk performance/quick partition (bytes passed per second)
- **Embed:** Embeds the windows of the useless app on the panel
- **Eyes:** The eyes have examined you
- **Fsguard:** Scans the selected free disk space
- **Genmon:** Displays the script/program, captures the output (stdout), and displays the resulting thread in the panel
- **Reference:** It is a small plugin written by Mark Trompell to consistently display information from various applications

- **Mail Clock:** Multi-thread, multi-mail box, multi-protocol plugin to check emails from time to time

- **Mount:** It is a panel mounting/lowering utility

- **MPC:** MPD client plugin, Music Player Daemon

- **Netload:** Shows the current load of social media, currently running on Linux, * BSD, Sun Solaris, HP_UX, and macOS X

- **Notes:** Provides sticky notes for your computer

- **Locations:** It can access the folders, documents, and removable media menu

- **PulseAudio Plugin:** Adjusts audio volume and controls media players on xfce desktop

- **Sample:** Sample plugin developers can use as a basis for new panel plugins

- **Sensors:** Hardware Panel Hardware Plugin.

- **Smart Bookmark:** Allows you to do direct online searches for sites like google or Debian Bugzilla

- **Statusnotifier:** Provides a panel location for status notification items (app references)

- **Stopwatch:** Keeps track of the past

- **Systemload:** Displays current CPU load, working memory, switching space, and system downtime

- **Time:** This allows the user to activate the alarm at a specific time or end the calculation time

- **Time Is Up:** Take a break from the computer every X minutes

- **Verve:** The free command line plugin for the xfce panel

- **Wavelan:** Displays statistics from the LAN wireless interface (signal status, signal quality, network name (SSID)). Supports NetBSD, OpenBSD, FreeBSD, and Linux

- **Weather:** Displays current temperature and weather, using weather data provided by xoap.weather.com

- **Whiskermenu:** A menu that provides access to favorites, recently used, and search applications installed

- **Xkb:** Sets and uses multiple (currently up to 4 due to X11 protocol limit) keyboard layouts

xfdesktop Usage – Desktop Manager

The xdesktop is responsible for drawing the desktop's background, or "wallpaper," and the images on the desktop.

Background

The background is made of solid color or gradient, with an image (optional) mounted on top. Images can be used in various ways to complement the screen, such as enlargement, zoom, scale, center, or tilt. Instead of a picture, a color or gradient of two colors can be used as a desktop background. Gradients can be horizontal or vertical.

Since the back end can be color-coded and the image at the top, you can mix both to produce stimulating effects using images that show across the various colors. If the image is medium or scaled and does not fill in the correct size, the background color is displayed on both sides of the image.

Thumbnails (Icons)

The icons drawn on the desktop can be file/launcher icons or minimized windows applications. It can be customized to various sizes, with custom font and tooltip size, with the option to turn off tooltips completely. The file/launcher option displays files, folders, and launchers as icons on your desktop that can be edited, composed, copied, and exported, modeled to have the same look and feel as a regular file manager. Launches can be configured with the right-click and select Edit Launcher.

Note: You can use the custom folder icon optionally to use external images in the desktop icon folders.

The reduced application icons option displays minimized windows as icons on your desktop and can be edited freely, and provides window controls such as a drop-down menu when you click the program icon to the left of the title bar with the right-click.

All icons can be found in the ~ / Desktop directory and accessed via the file manager. Launches are desktop files that can be edited using a text editor or created and placed in the ~ / Desktop directory to create application icons for desktops.

xfce4-session – Session Manager

xfce4-session is the xfce session manager. Its job is to save the status of your desktop (open applications and their location) and restore it during the next launch. You can create several different sessions and choose one of the first.

1. **Show selector on the login:** Once set, the session manager will ask you to select a session each time you log in to xfce.

2. **Automatically save session on the checkout:** This option instructs the session manager to save the current session automatically when you log out.

3. **Notify on exit:** This option disables the checkbox exit. Whether the time will be saved or not depends on whether you enable automatic saving of exit sessions or not.

4. **Kiosk mode:** Session Manager provides Kiosk mode support, which helps prevent others from making changes to their session settings. To use it you have to edit or create a $ {sysconfdir} / xdg / xfce4 / kiosk / kioskrc file.

The way to define the format of this file is to use an example. The xfce4-session section of your kioskrc file may look like this:

```
[xfce4-session]
CustomizeSplash = ALL
CustomizeChooser = ALL
CustomizeLogout = ALL
CustomizeCompatibility =% wheel
Turn off =% wheel
CustomizeSecurity = NO
```

It allows all users to change their splash, selector, and exit settings, but only allows users in the group wheel to customize the interaction and shutdown settings. No one will be allowed to adjust the security settings.

The session manager supports the following KIOSK capabilities:

• **Customize Splash:** The user to allowed to customize the splash screen.

- **CustomChooser:** It allows the user to customize the session selector settings.

- **CustomizeGet Out:** Whether the user is allowed or not to customize the opt-out settings.

- **CustomizeCompatibility:** Whether the user is allowed to customize the interaction settings (KDE/Gnome compat).

- **CustomizeSecurity:** Whether the user is allowed or not to make security settings. It is one of the MOST basic settings as it prevents users (actually libICE) from tying the TCP port.

- **Shutdown:** Whether the user can shutdown (restart or shutdown) the system, the restart options and the power off in the shutdown dialog will be gray if the user does not have this capability.

ConsoleKit

xfce Desktop requires a valid ConsoleKit session for it to work correctly. ConsoleKit is used for system actions such as shutdown, pause, and (un) mounting devices (via Polkit for authorization). If you are using a desktop manager, DM will take care of that. For the console to start, this is slightly different.

When entering the console (without starting X), use the ck-list-sessions. Make sure it works = TRUE and local = TRUE. If not, check that the consolekit library is loaded with PAM.

At 4.10, start xfce with startxfce4 --with-ck-launch. It will start the xfce4-session with the ck-launch-session. At 4.10, the xfce4-session will oversee the launch of the dbus session.

Once xfce is launched, and you use ck-list-sessions, a second session is created that should give you all the necessary permissions to mount devices and use power actions (depending on each distribution, you may need to add them to different user groups).

Autostart Settings App

The applications can start automatically based on specific triggers, e.g., start a session, log out. You can add or remove apps. In addition, all applications with their desktop files stored in the user interface can be edited. When a saved session is restored, all resources from that backup session will be started in addition to the items selected in this list. Everything in italic font style is for a different desktop, but you can still enable it.

You can select when to start the action chosen in the drop-down menu. This allows you to use custom commands, e.g., exit, pause.

Thunar File Manager

It is the modern file manager of xfce Desktop Environment. Thunar is built from the ground up to be quick and easy to use. It is clean and intuitive and does not automatically include confusing or useless options. Thunar starts fast, and navigation in files and folders is fast and responsive.

- **File Manager Window:** Works with Thunar window and configures the layout.

- **Working with Files and Folders:** It is an essential file management.

- **Using Removable Media:** It works with removable devices.

- **File Management Preferences:** It is detailed for different file manager options.

- **Frequently Asked Questions:** Some of the spontaneous things you might want to know.

- **Advanced Settings:** xfce4-settings editor, customize gtk3 CSS and other improvements.

- **UNIX File System:** To understand the UNIX File System.

Thunar plugins can be installed as separate packages and extend Thunar functionality. Most Thunar plugins provide additional options for files in the context menu or shortcut. The interface used by thunar plugins is the thunarx API.

- **Bulk Renamer:** It can rename multiple files at once.

- **Custom Actions:** It can custom commands associated with common mime types or extensions.

- **Archive (thunar-archive plugin):** Allows you to create and extract archive files.

- **Media Tags Plugin (thunar-media-tags-plugin):** It enhances support for ID3 tags.

- **Share Plugin (thunar-shares-plugin):** It can quickly share folders using Samba without requiring root access.

- **Volume Manager (thunar-volman):** It can improve automatic management of removable Thunar devices.

- **VCS Plugin (thunar-vcs-plugin):** It can add conversion actions and GIT to the context menu.

xfce4-terminal – Terminal Emulator

The UNIX operating system was designed as a text-only program, controlled by keyboard input instructions. It is known as the command-line interface (CLI) interface. X Window System, xfce, and other projects have added a graphical user interface to UNIX, which you use now. But adding a user interface does not mean CLI is dead. CLI still exists and is often the easiest, fastest, and most powerful way to do a particular job.

A terminal is a so-called X terminal emulator, commonly referred to as a terminal or shell. It provides the same as the old text screen on your desktop, but the one that can easily share the screen with other graphic apps. Windows users may already be familiar with the MS-DOS Prompt service, which has the same function of providing a DOS command-line under Windows, although one should be aware that UNIX CLI offers ease of use than DOS.

The term mimics the xterm system developed by X Consortium. Next, the xterm application mimics the DEC VT102 terminal and supports the DEC VT220 escape sequence. Terminal accepts all exit programs used by VT102 and VT220 terminals for functions such as cursor setting and screen clearing.

xfce Terminal is a lightweight and easy-to-use terminal emulator app with many advanced features, including drag, tabs, unlimited scrolling, full colors, fonts, and transparent background.

What Makes the Terminal So Special?

Improved terminal features include a simple configuration interface, the ability to use multiple tabs with terminals in a single window, the possibility of having a non-transparent terminal, and integrated mode. It helps you save space on your desktop.

The following key features are available:

- Multiple tabs per window

- Custom toolbars, which can be customized using the toolbar editor with integrated images

- Ability to configure almost all aspects of the Terminal in the Favorites dialog and a host of so-called hidden options

- Good integration with xfce desktop space in particular, but also with every other Linux desktop

- Session management support

- Real multihead support (both MultiScreen and Xinerama mode)

- Compliance (see Freedesktop.org website)

- DBus-based service center to minimize overall service usage

- High level of savings by making good use of GTK + and GObject.

- Apart from these important features, Terminal supports all the features you expect from a modern terminal character.

xfce4-appfinder – Program Finder

Application Finder is a program for finding and launching applications installed on your system and then issuing commands immediately. It does this by searching your .desktop file system and displaying a fixed list of all GUI applications in your system.

Application Finder has two options:

1. **Folder Mode:** You can directly search for commands and run them when Enter is pressed.

2. **Extended Mode:** You can search for applications installed on your system and view command history.

To switch between modes, you can click the Up or Down arrow keys when the entry is fixed or click the arrow on the right-hand side of the input.

Collapsed Run Mode
The Folder View Finder shows only search entries, selected program icons, close buttons, and then Launch. As you type entries, it will search for available applications, command history, and custom actions, showing the same potential in the pop-up menu. If one match is found, it will be automatically terminated at login.

Expanded Search Mode

In Extended Views, the included search function acts as the currently selected category filter. When Enter is pressed, a highlighted item is introduced.

xfconf – Storage System

Xfconf is a consecutive setup system (similar to a tree) in which baby nodes are close to the root called "channels." All settings under channel nodes are called "properties."

Valid channel names and locations are made up of uppercase letters ASCII US-English A–Z, lowercase letters a–z, numbers 0–9, dash (–), and underscore (_). No other characters are allowed in channel names. Letters less than (()) and larger than (()) (also called "angle brackets") are also allowed in property names but not in channel names.

Property names refer to the "full route" under its channel, for example: "/main/history-window/last-accessed." Of course, when inquiring about a particular property, the channel should be specified separately.

xfce4-dev-tools – Development Tools

These tools are a collection of tools and tools for xfce developers and people who want to build xfce from Git. Additionally, it contains an xfce developer handbook.

XFCE CI TEMPLATES

[CI Folder] (ci /) contains a build_project.yml template for building various xfce projects, as well as supporting documents such as "build_libs. sh," which handles building any dependencies needed. It helps us to avoid repeating the same building code for each project.

Exo – Assistant Applications

Exo is an expandable library used on the xfce desktop, first developed by oscillation. It has auxiliary applications utilized throughout the xfce desktop to manage selected applications and edit .desktop files. Exo is focused on application development and is considered suitable for use in production. You no longer need to define the previous EXO_API_SUBJECT_ TO_CHANGE logo.

```
Desktop Object Editor (exo-desktop-item-edit) -
      .desktop file editor
```

Exo-preferred-applications – Select the default applications used for various services, i.e., the web browser used to open links.

Exo – exo-desktop-item-edit – Desktop Object Editor
Exo-desktop-item-edit allows the user to easily create and modify application launches or .desktop files in the GUI.

Creating a Launcher
To create a new launcher, by terminal:

```
exo-desktop-item-edit --create-new [DIRECTORY]
Right-click the space on the desktop and select Create
    Launcher to create a launcher on the desktop.
```

Editing Launches
To edit existing launchers on the desktop, right-click the launcher you want to edit and select the Edit Startup menu. When the launcher is on the panel, right-click the launcher you want and select Properties. Then, in the Startup window that appears, select Edit the currently selected item in the General tab.

The edit/create window lets you customize various features of your launcher or desktop file. Both editor and creator contain the same fields to customize your file.

- **Name:** Application launcher title.

- **Note:** Additional text is displayed when you hover over the launcher on the desktop or panel. This text can also be displayed under the icon and title in other apps as an app finder.

- **Command:** It is the command that is used when launching the launcher.

- **Task List:** You can set up a custom performance guide for the startup program.

- **Thumbnail:** The custom icon can be used for your launcher, and you can choose from a list of applications, system icons, or an image file.

- **Use startup notification:** It enables startup notification when launching.

- **Run to the terminal:** To enable this, use the terminal window command where the launcher is used.

Exo – exo-preferred-applications – Preferred Applications

From xfce 4.3, a popular application framework can easily manage the default applications used for various services. It includes selecting the default application like a web browser, mail reader, file manager, and terminal template. You can access any app by clicking the Preferences Request button in the xfce Settings Manager. The window contains two tabs containing various applications that can be customized.

Exo-preferred-applications support the following types of applications:

- **Web Browser:** Popular Web Browser

- **MailReader:** Preferred Mail Reader

- **FileManager:** Popular File Manager

- **TerminalEmulator:** Preferred Terminal Emulator

xfce4-settings – Settings Manager

It is an application that contains a link to all xfce settings dialogs. Some of these conversations will be embedded in the Settings Manager window. If that does not match, the Settings Manager will display scroll bars to access the hidden parts of the dialog.

The Internet

The pre-selected tab window for applications allows you to select your default web browser and Mail Reader. The Web Browser is used to open links and display texts, while Mail Reader will compose emails when users click on email addresses, for example, on a website.

Preferred Internet Applications

To select a different web browser that will be used to open links, click the drop-down in the Web Browser, and choose your selected browser from the list of available web browsers.

If the browser you are looking for cannot be found by the system automatically, you can select More in the drop-down menu, and a box will appear asking you to enter a customizable browser command.

Resources

The second Preferences window tab lets you select your default File Manager and Final Template. The File Manager has used the system to open folders to browse files, and the preferred Terminal Emulator will be used to run documents and applications that need to be run in the CLI area. For custom terminal commands, "% s" is in the command. If not, the same rules apply as described in the paragraph above.

Command-Line Options

A simple command line in front of the end of the popular apps framework is included, called exo-open. You can use this application to launch the preferred application for a particular category or open URLs with the default URL holder.

Exo-open supports two modes of operation. The first one will automatically launch a popular application for a particular category by selecting the pass parameter. For example, to launch a command mutt in a set of Airport Emulator set, you can use:

```
exo-open - launch TerminalEmulator mutt
```

To open the popular Web Browser without additional restrictions, the following command will be used:

```
exo-open - launch WebBrowser
```

The second exo-supported mode opens all the parameters with the default URL holders. Here URLs refer to fully qualified URLs (i.e., https://www.xfce.org/ or mailto: xfce4-dev@xfce.org), or local methods (i.e., ~ / myfile.txt). For example, to open ~ / file1.mp3 and ~ / file2.txt in the default settings, you will use the following command:

```
exo-open ~ / file1.mp3 ~ / file2.txt
```

To start composing an email list for xfce4-dev addresses in the preferred email reading application, for example, you can type:

```
exo-open mailto: xfce4-dev@xfce.org
```

Required packages:

- GIO
- GLib
- GTK+
- gthread

- libxfce4ui

- libxfce4util

- gio-unix

Garcon – Menu Library

Garcon is the implementation of a specific menu specific to Freedesktop.org. It is based on GLib/GIO and aims to cover all details except asset menus. It was started as a complete redesign of the previous xfce menu library called libxfce4menu, which, unlike garcon, lacked the essential integration of menu loading menu menus.

Garcon incorporates almost every part of the menu specificity except the asset menus and a few XML attributes. Unlike libxfce4menu, it can also load customized menus and menu editors like Alacarte, as menu integration is now supported. The only significant feature that is still missing is monitoring menus and menu items to make changes. It is something that will work on in the next release. The garcon API will probably not freeze until it is released at 1.0.0 (currently version 0.7.1)!

Tumbler – Thumbnail Service

Tumbler is a DBus service for requesting icons for various URI programs and MIME types. The DBus icon management is described below. The DBus icon management specification is in the standard DBus API to handle icon creation.

With DBus specification for thumbnail management, applications do not have to use it to manage thumbnails themselves. When an icon is found, they can send the icon function to a special service. The service then dials when it has finished producing the thumbnail.

This service has benefits:

- There is no need to link sophisticated software pieces to all applications that work with icons.

- The possibilities for closed format icons (which may be patented) coexist with free desktop software (just) that wants to display files as thumbnails.

- Reuse of existing infrastructure instead of having all the applications redesigned.

- The complexity of the LIFO line, the setting of I/O, and the arrangement of small background images are no longer the function of the application developer.

It is written in an object-oriented manner using GLib and GObject. Its modular structure makes it very flexible and valuable in most situations. Provides plugin workspaces for extending URI schemes and MIME types for which icons can be created and changes the background storage used to store icons on the disk. Tumbler functionality can be extended using special icon services used in conjunction with the D-Bus icon management. Tumbler is used by Thunar, Ristretto, and xfce.

Reasons to Use xfce as Your Desktop
Simplicity
xfce has become very similar to simplicity over the years. To put it bluntly, "simple" here does not mean little. It means that doing things is very easy. Users will not have to look long to find what they are looking for. For example, in my opinion, the Settings Manager is the best among all the other major DEs. Straight and well divided. Like this one example, all other parts of xfce are well-designed.

Lightweight Nature
Another famous and essential feature of xfce is its lightweight nature. In most cases, xfce is a small package and convenient for system resources. For example, in idle mode, xfce uses just ~ 400 Mb of RAM in my system. Firefox open, and 1080p video playback has expanded to 1.20 GB.

Performance
Now, because of all that has been said before, it translates into the fact that xfce works well on most hardware. It is fast and responsive. After testing and knowledge, users can quickly navigate the system and use high productivity. CPU usage in idle (or low) mode is shallow.

Modularity
The xfce architecture is entirely modular. Most of the DE components are different, making the system more flexible. For example, the default window manager is Xfwm, the setup program is Xfconf, the desktop manager is Xfdesktop, and the session manager is xfce4-session. These items are

usually integrated and do not change in most DEs. Settings for all components are integrated into the Settings Manager to provide a compact visual interface for convenience.

Configuration

xfce has an excellent range of configurations. From visual interface to hardware preferences, everything can be easily adjusted. The Settings Manager covers almost everything needed. In terms of appearance, there are websites like these available for you to find various icon themes and system themes.

Stability

xfce project is stable. There are a few bugs in it, and they are often problematic because of the hardened drivers. The use of the system is entirely smooth. The review cycle is significantly stable. xfce 4.14 was just released after almost 4.5 years of development! All DE components work very well to provide smooth information to users.

Panel

One of the most distinctive features of the xfce panel. It is highly customizable, with several plugins provided by the xfce team. The width and length of the panel can be adjusted, and users can get it on the screen as they wish. You can also place the panel in the center of the display if you want. You can add as many panels as you like. By default, one of the xfce panels has already provided them installed as a dock at the bottom of the screen. Other settings are provided, such as bar blurring, and thumbnail size adjustment.

Plugins

The built-in plugins are excellent. They include many applications and requirements. There are various plugins in xfce as listed below.

- **CPU Graph Plugin**: This shows graphical CPU usage right on the panel. It helps to track what type of program is included in your system. We find looking at this plugin whenever we check out a new program, new distribution based on xfce, or different hardware than usual.

- **Pomodoro Plugin**: This is an excellent production plugin. The Pomodoro method is widely used to increase human productivity. It is followed by a pattern of job exchanges and periods of rest, for example, 25-minute workout, 5-minute break. xfce provides a plugin to keep track of time when using this method. It brings the Pomodoro timer to the panel timer. This plugin belongs to a third party. The link can be found here, along with the installation instructions.

- **Mail Watcher Plugin**: This can be used to receive email notifications from multiple services in one place.

- **Notes Plugin**: This is not a very unusual feature, but the notes panel plugins make it accessible at another level. Creating different groups of notes and customizing the notes app makes it even more helpful.

- **Verve Command-Line Plugin**: This is a helpful feature in some way. Using this seemingly empty line, users can open applications, go to websites, search using DuckDuckGo, open emails, and open the directory completely.

- **Workspace Plugin**: Workplaces are an excellent production feature on desktops. xfce was one of the first DEs to adopt this feature, so this feature is very mature in xfce. You can see the workspaces right on the panel for better access, with a small view of the open windows throughout.

Integration

Every program you put on xfce looks excellent no matter what the original DE was developed for or whether the image enhancement kit was Qt or GTK. GTK, Qt, GNOME, Electron, etc., everything seems to go into the system. All GTK systems work flawlessly, but Qt systems have minor problems.

Easy to Fix

Using an operating system will break down once and for all unless you are incredibly familiar with the stable distribution, maybe because of a review or a user's error. In some DEs, many things are integrated into each other. If you violate the application menu, you will probably need to reset the entire DE again. In xfce, as we have said, things are modular. If the menu breaks, re-enter that! It is straightforward to do.

xfce Components

We are now finally moving out of the standard features of the DE itself and into the software components you will be using. Software applications are part of DE itself, and when we talk about DE, we have to reconsider. We will do this through some of them, one by one.

- **Thunar (File Manager):** xfce's default file manager is Thunar, a program you know. It has excellent simple features like viewing items in list/icon mode, resizing objects, shortcut panel in other drives, partitions or folders, tab functionality, etc. It is stable and has all the basic features of users.

- **Mousepad (Text Editor):** Mousepad is fast and gets the job done. There are essential features for changing color theme, font style, search, etc. Mousepad has some features that make it suitable for editing. Granted, it is not IDE, but some basic requirements have been considered. There is syntax highlighting, auto-zoom, line wrap, highlighting brackets, adjusting line numbers, adjustable tab size, rare and excellent features, etc. These features make it an excellent program solution, terminal included for compiling/creating program documents.

XFCE INSTALLATION

Since 1996, xfce Desktop has empowered users to have a graphical user interface (GUI), visually transforming your Linux server into a desktop-like environment for your desktop computer. With its simple appearance, xfce is lightweight in server hardware and faster than GNOME and KDE launches. Once you are done with this bit of tutorial, you will be able to share and connect to the xfce GUI by moving on to the following tutorial on installing VNC. These commands are intended to install the xfce Desktop Environment on the Ubuntu 16.04 LTS server. Logged in as root user, but for rootless users, it precedes all commands with the word sudo.

Stepwise xfce Installation

Step 1: Update the apt-get

Thinking of the best practices, we will review them before proceeding with the installation of xfce 4.

```
$ apt-get update
```

Step 2: Install xfce4 desktop environment

With one command, we can install xfce itself and other valuable resources that come with xfce:

```
$ sudo apt-get install -y xfce4 xfce4-goodies
Or
$ sudo apt-get install -y xfce4
```

After installation you will see the first window of xfce Desktop environment:

xfce Desktop Environment.

xfce Removal

Step 1:

Run each of these commands so that the apt-get can use it while cleaning xfce.

```
$ sudo apt-get -f install
$ sudo apt-get clean
$ sudo apt-get autoclean
$ sudo apt-get update
```

Step 2:

Clean xfce on your Ubuntu server:
```
$ sudo apt-get purge xfce4
```

Products and Distribution Using xfce

xfce has been included as one of the most iconic workshops in the Pandora handheld game.

It is the default desktop location for the following Linux distributions:

- BackBox
- Black Lab Linux
- Uninstall GNU + Linux OS
- Dragora GNU/Linux-free
- Buntus
- EndeavourOS
- GalliumOS
- GhostBSD community program
- Kali Linux
- Linux Lite
- Mananjaro
- MX Linux
- Mythbuntu
- QubesOS
- SalineOS
- SolydXK (SolydX)
- SystemRescueCD
- UberStudent
- Ubuntu Studio (Up to 20.04)

- Xebian

- Xevuan

- ubuntu

It is also a standard desktop option in FreeBSD and many other Linux distributors are not listed above such as Arch Linux, Debian, Ubuntu, openSUSE, Fedora, Kali, Korora, Linux Mint, Slackware, Mageia, OpenMandriva, Void Linux, and the Zorin OS. Kali Linux also uses xfce as a desktop space where it works on the ARM platform. Debian makes a unique netinstall CD available, including xfce as the default desktop. In 2013, Debian briefly made the default location, replacing GNOME.

Compatibility

xfce4 is written using GTK + 2 and is fully compliant with GNOME 2 applications. Desktop and window manager (xfwm4) and compatible with K Desktop Environment (KDE), Gnome, and Rox-Filer session/file manager.

Advantages

- xfce is a simple structure. It has very little memory and less CPU usage than KDE or GNOME.

- The main advantage that follows is the simplicity of the visual space. xfce desktop is not complete and straightforward. The primary desktop has two panels with a straight line of icons. One panel has a basic app launcher, and the other has a workspace switch.

- xfce Default File Manager is simple, easy to use, and optimized with a nearly flat reading curve.

- The xfce desktop is unusually stable with regular updates, with new releases occurring over a three-year cycle.

- xfce4 has a powerful screen where tabs allow multiple terminals in a single window.

- Lastly, xfce has a high level of flexibility (below KDE but much higher than GNOME) and thus allows users to enjoy moderation.

Disadvantages

- In some cases, the formation of a small weight may result in a bottle when the data is processed in a large area, and the area itself does not support power.

- One special case for using other packages needs to be installed to make the disk work.

- It does not have the advanced features that GNOME videos have.

- A tool to speed up or slow down media playback is not possible in xfce.

CHAPTER SUMMARY

In this chapter, we have covered an introduction of xfce with its features, history, core projects, applications, and development. Also, we have a separate section where you get a history of xfce versions.

CHAPTER 5

MATE Desktop Environment

IN THIS CHAPTER

- ➤ Introduction
- ➤ History of Mate
- ➤ Version
- ➤ Installing MATE in Ubuntu
- ➤ MATE applications
- ➤ Installation on various OS

In the previous chapter, we discussed xfce. In this chapter, we will learn about the working of the MATE Desktop Environment, and its installation and features.

INTRODUCTION

MATE is a free and open-source desktop application running Linux, BSD, and illumos applications. The desktop area contains everything you see on your computer screen. Panels open applications, display notifications, and display time. It can control your windows and allows you to switch between them easily.

DOI: 10.1201/9781003308676-5

When you look at the screenshot, all the elements make you think of Windows as Windows and macOS as macOS. Windows and macOS both provide one desktop space. Significant changes surrounding the new Windows release are often related to the desktop area, such as removing or adding Start Menu and changes to the theme. Windows 8 had more connectivity provided on touch screens than desktops. People who did not like the change could not upgrade to Windows 8 if they wanted to maintain the visual interface they were comfortable with.

For Linux, this is not a problem. You can switch to another desktop and continue using the latest Linux software. And if your preferred visual interface is about to undergo a significant change, MATE is an example that doesn't mean you have to say goodbye to what you know.

HISTORY OF MATE

MATE is based on GNOME, the most popular free open-source desktop software like Linux. However, to say that MATE is based on GNOME is an understatement. MATE was born as a continuation of GNOME 2 after GNOME 3 in 2011.

The split is because GNOME 3 introduced a new interface called GNOME Shell, based on a standard desktop design. Since the project was open-source, developers unhappy with the change were free to take the existing GNOME 2 code and continue working on that instead. Doing this is called a "break" project. An Arch Linux user called Perberos has started a MATE project, and others are jumping on board right away.

Although many Linux applications have adopted GNOME 3, MATE has attracted more users over the past decade. Some started using Linux many years after they broke up with GNOME. That means they are using MATE for non-transformative reasons. Few consider it one of Linux's most stable and flexible features. MATE is a continuation of GNOME 2. It provides an attractive desktop environment using common Linux metaphors and other applications such as Unix. It is under active development to support new technologies while maintaining the everyday desktop experience.

VERSIONS

Here we are going to discuss the versions of MATE, as given below.

- MATE version 1.18

- MATE version 1.20

- MATE version 1.24

- MATE version 1.26

So the first version of MATE is given below.

MATE VERSION 1.18

After six months of development, the MATE Desktop team is proud to announce the release of MATE Desktop 1.18. We want to thank every MATE donor for his help in making this release possible. The release focuses on completing the move to GTK3 + and implementing new technologies that will replace the now-defunct MATE Desktop 1.16 reliability components.

Features

1. The entire MATE Desktop suite for apps and components is now GTK3 + only.

2. They added mouse libinput support, a touchpad, different hand-holding settings (left/right), and speed/limit acceleration.

4. It has highly improved accessibility support (especially for visually impaired users).

5. The lock screen will load the users' selected background instead of the default system-defined default.

6. The MATE panel receives several improvements, including:

   - Added desktop support for more startup options

   - Added StatusNotifier support

   - Additional support for the Menubar menu editor, if installed, is preferred over Mozo

7. Engram, archive manager, includes a few improvements:

   - Added ears and war to the list of supported types.

   - It can check the rar/unrar version if the correct date is shown.

   - Stable rar compress and 7z are divided into volumes.

8. Notifications now use support for action icons, for example, play control icons used by media players.

9. Reset font viewer to add font browsing mode, support TTC fonts, and expose to menus automatically.

10. Replaced UPower reduced stop/hibernate functions with ConsoleKit2 equivalent.

11. MATE terminal adds power to close tabs with the middle mouse button.

12. Atril, the document viewer, has the most advanced page load times and adds support for extracting history used by other jokes.

13. Many retractable GTK + modes have been modified, and many bugs have been fixed.

With all C and Python plugins, Plugin systems in Eye of MATE and Pluma are included in libpeas.

MATE VERSION 1.20

After 11 months, the MATE 1.18 team is excited to announce the release of MATE Desktop 1.20. The theme of this release was stabilizing the MATE Desktop by changing the decrypted code and modernizing most of the codebase. We also upgraded our window manager (Marco) and added HiDPI support. Along the way, we fixed hundreds of bugs.

The title changes to MATE Desktop 1.20 are:

- MATE Desktop 1.20 supports HiDPI indicators with dynamic detection and measurement.

- Marco now supports DRI3 and Xpresent, if any.

- Now Marco supports drag-and-drop quadrant window tile, cursor keys can be used to navigate the Alt + Tab switch, and keyboard shortcuts to move windows to another monitor have been added.

- Support for Global Menu providers as a close-panel-app menu has been added.

- The MATE panel has the most advanced Status Identification (SNI) support.

- Bookmarks now support GTK3 + locations.

- MATE Terminal now supports background images, adding Solarized themes and combining keys to switch tabs.

- The Invest applet has been released from MATE Applets.

- MATE Themes has seen significant improvements to the full use of all style classes revealed by GTK 3.22.

- Engrampa, an archive viewer, has developed support for 7z encrypted archives.

- The MATE Sensors Applet (finally) supports disks2.

- OpenBSD authentication is supported in MATE Screensaver, and minizip in April supports additional BSD variants.

MATE VERSION 1.24

After a year, the MATE team released MATE Desktop 1.224. This release contains a lot of key bug fixes and under-hood upgrades.

The MATE panel got a lot of crazy work to work with the Wayland display backend.

- The display applet received a full update, allowing better control of direct monitoring from the panel.

- The timer applet now has the best mouse interaction.

- Wanda the Fish now works well on HiDPI displays.

After a year of development, the MATE Desktop team released MATE 1.24.

This release contains many new features, bug fixes, and general improvements. Some of the most important features include:

- Engrampa now has support for several additional formats and consistent support for passwords and Unicode characters in some of them.

- Eye of MATE now has Wayland support and has added support for color-embedded profiles.

- Icon creation has been redesigned and modified in a few places.

- Additional support for webp files.

- The Control Center now displays its icons correctly on the HiDPI displays.

- The new Time and Date Manager app has been added.

- The Mouse app now supports acceleration profiles.

- The Favorite App application has been upgraded for accessibility, with better support for integration with IM clients.

- Indicator Applet has slightly better interaction with icons of unusual size.

- Speaking of icons, the network manager applet icons in our themes have been completely redesigned and can now be enjoyed on HiDPI mirrors.

- The notification daemon now supports Do Not Disturb mode.

- The MATE panel had a few bugs that caused crashes in the past when it changed buildings.

- Mozo, the menu editor, now supports Undo and Redo actions.

- Pluma plugins have now completely changed to Python 3.

- Pluma no longer has to be jealous of other sophisticated editors, as it can now show formatting signs.

MATE VERSION 1.26

A year later, the MATE team released MATE Desktop 1.24. This release contains several key bug fixes and under-hood upgrades. The title changes to MATE Desktop 1.26 are:

1. Added Wayland support for Atril, System Monitor, Pluma, Terminal, and other Desktop components. In addition, the Counter and Terminal can now be built with Meson.

2. Pluma has been under significant development.

   - A small new map gives you a quick overview of the content.

   - A new grid background pattern transforms Pluma into a writing pad.

- The filter plugin now supports retrieval actions.

- Add a shortcut to the display/hide line numbers, Ctrl + Y.

- Favorite dialog needs to be redesigned to fit all new features.

3. With the new Pluma Plugins, our text editor can be converted into a rich IDE with features that support Bracket completion, comment/comment code, built-in terminal, and Word Completion.

4. The Control Center has an improved Windows preferences box with more options. In the Display, Box adds the display rating option.

5. Notifications were even more helpful as they received link support. And finally, they added the Do Not Disturb applet.

6. Speaking of applets, the Window List applet has a new option to disable the mouse scrolling. The window icons of your choice are now clearer and clearer as they are translated as cairo locations.

7. The Netspeed applet automatically displays additional information and gets netlink support.

8. The calculator is included in the GNU MPFR/MPC library, which means more precision, faster calculation, and more functions.

9. Window manager Marco now restores minimized windows to their original, more reliable location.

10. Engram received support for Electronic Publication (EPUB) and ARC archives. Additionally, it can now open encrypted RAR archives.

11. Power Manager has a new option to allow keyboard blurring and moved to libsecret.

INSTALLING MATE IN UBUNTU

MATE is available in official libraries and may be included in any of the following: The collaborator group contains the main desktop area required for general MATE information. The mate-extra team contains additional resources and applications that integrate well with the MATE desktop. Just adding a mate-extra team will not draw the whole group of partners depending on it. If users want to install all MATE packages, you must install both groups. The base desktop contains marco, mate-panel, and mate-session-manager.

Terms for the Official Repository

The software repository is the last place where software packages are downloaded for installation. The official archives of Arch Linux contain essential and popular software, easily accessible via pacman. Package maintainers store them. Packages in official repositories are constantly updated: when a package is upgraded, its old version is removed from the repository. There are no major Arch releases: each package is upgraded as new versions are available from mounting sources. Each repository is always compatible, i.e., the packages it carries always have compatible versions.

Installing MATE

The tasksel package will be used to install MATE on our Ubuntu systems.

- **Installing Tasksel**
    - Tasksel is an Ubuntu package that provides an interface that allows users to install packages on their systems as if they were performing a specific task. To use tasksel, we first need to install it on our systems.
    - To do this, open the terminal by either hitting the Ctrl + Alt + T keys or use the dash to get access to the list of all applications installed. After opening the terminal, enter the following command:

    ```
    $ sudo apt install tasksel
    ```

    - To verify whether tasksel has been installed, enter the following command into the terminal:

    ```
    $ sudo tasksel
    ```

    - If you see a screen similar to the one shown below, then the tasksel has been installed onto your system. Press Esc to go back to the terminal.

- **Installing MATE**
    - Once the tasksel has been installed, our next step is to install the MATE Desktop Environment on our Ubuntu system. There are two versions of Plasma that are available for installation: minimal and full.
    - The minimal version comes only with the MATE desktop environment. No other applications are installed, and users can install whichever they want later on. This version is quite useful if users don't want to use too much of their memory or if users want to stick to the default Ubuntu applications.

- To install this version, enter the following command into the terminal:

```
$ sudo tasksel install ubuntu-mate-desktop
OR
$ sudo tasksel install ubuntu-mate-core
```

- During installation, it will display a prompt asking you to configure sddm, which is the display manager for MATE.

- After complete installation, you need to restart your system then login screen appears.

- Enter your username and password to log in to your system, and a black screen will appear.

- The first screen of MATE will be as,

MATE Desktop.

MATE APPLICATIONS

MATE is largely built with GNOME 2 applications and applications, forged and renamed to avoid conflicts with their GNOME 3 counterparts. Below is a list of standard GNOME applications renamed with MATE.

Application	MATE
File manager	caja
Menu editor	mozo
Window manager	marco
Text editor	pluma
Image viewer	Eye of MATE (eom)
Document viewer	atril
Archive manager	engrampa

Some applications and key features pre-installed with GNOME (such as GNOME Terminal, GNOME Panel, GNOME Menus, etc.) have changed the start to MATE to MATE Panel, MATE Menus, etc.

CORE APPLICATIONS

MATE has cracked down on several applications, such as GNOME Core Applications, and developers have written a few more applications from scratch. Fork applications have new names, many of them Spanish. MATE applications include:

Caja

In Spanish, Caja, which means "box," is a file manager embedded in the GNOME Files, formerly known as Nautilus. Caja serves as a natural component of the MATE desktop. Since the Caja, a fork has been developed and new features added.

Features

Caja has many functions: creating a folder and document, displaying files and folders, searching and managing files, and installing or extracting fonts. Caja can manage different types of file formats.

Caja added improvements to the latest release of MATE 1.26. Some of these are:

- In Caja, users can select the new Bookmarks sidebar.

- Caja received drive formatting support from the context menu.

Caja also offers many extensions such as:

- **Python:** caja-rename

- **Python:** caja-admin

- **libcaja:** seahorse

- atril

- episode

Caja is a well-known file manager at Linux. The Caja text extension allows users to add texts easily.

Pluma

It is the free and open-source default text editor for the MATE desktop Linux distribution. Pluma is an image application that supports editing multiple text files in a single window. It fully supports international text using its Unicode UTF-8 encoding. As a general-purpose text editor, Pluma supports many common editor features and emphasizes simplicity and ease of use. Its key feature set includes syntax highlighting source code, automatic retransmission, and print support with print preview.

Features

Pluma includes the full integration of MATE, which provides for dragging and dropping to and from Caja (MATE file manager), application of the MATE help system, the MATE Visual File System, and the MATE print framework. It has Multiple Document Interface GUI tabs for organizing multiple files. Tabs can be moved between different windows by the user. It can edit remote files using GVfs. Supports complete overhaul and system overhaul and search and replacement. Other code-based features include line numbers, brackets matching, text wrapping, current line highlighting, auto-loading, and a backup copy of the file.

Pluma features include multilingual spell checking with Enchant and a flexible plugin system that allows the addition of new features, for example, captions and integration with external applications, including Python or Bash terminal. Many plugins are installed in Pluma itself, with many plugins in the pluma plugin package and online.

Pluma supports printing, including preview and print previews in PostScript and PDF files. Print options include text font, page size, layout, margins, optional printing of page titles and line numbers, and syntax highlighting.

List of Features

- Highlighting syntax

- Print and Print Preview Support

- File Restoration

- Full support for UTF-8 text

- Remote file editing support

- Search and Replace

- Adjustable plugin, with optional python support

- The perfect visual interface for favorites

- A small new map that gives you a quick overview of the content

- A new grid background pattern transforms Pluma into a writing pad

- The filter plugin supports retrieval actions

- Display shortcut/hide line numbers, Ctrl + Y

Atril

It is the official text viewer for the MATE Desktop Environment. A simple multi-page text viewer. It can display and print PostScript (PS), Encapsulated PostScript (EPS), DJVU, DVI, XPS, and portable document format files (PDF). The document also allows text search, copy to clipboard, hypertext navigation, and table bookmarks if supported by the document. Atril is Evince's fork.

Features

- Combined search that shows the number of results found and highlights results on the page

- Page icons

- **Documentation:** Once the reference information is included in the PDF, Evince will display it in tree format.

- **Printing Documents:** Using the GNOME/GTK print frame Evince can print any text I can open

- Viewed Encrypted Document

- International support

Supported Formats

- Portable Document Format (PDF) uses the Poppler backend

- Postscript uses GhostScript backend

- Multi-page tag file format (TIFF) file format

- Impressive

- DjVu uses the DjVuLibre backend

- DeVice Independent (DVI) – TeX software output format for editing

- Jokes

- Photos

Engrama

Engram is a MATE Environment archive manager resource. It's the Archives' Fork Manager. The software allows you to create and modify archives, archive content, view and edit archive content, and archive files. Engmpa is a free open-source software.

Supported File Types

- 7-Zip Compressed File (.7z)

- WinAce Compressed File (.ace)

- Compressed ALZip File (.alz)

- Archive with a small AIX (.ar) index

- ARC archive

- Compressed ARJ archive (.arj)

- Cabinet File (.cab)

- UNIX CPIO Archive (.cpio)

- Debian Linux Package (.deb) [Read Mode Only]

- ISO-9660 CD Disc Image (.iso) [Read Mode Only]

- Java archive (.jar)

- Java Enterprise (.ear) Archive

- Java (Archive) web archive

- LHA archive (.lzh, .lha)

- Depressed WinRAR archive (.rar)

- RAR Archived Comic Book (.cbr)

- RPM Linux Package (.rpm) [Read Mode Only]

Eye of MATE Image Viewer

The MATE Image Viewer Eye app lets you view single image files and large photo collections. You can launch the image viewer by double-clicking the image in your file manager or clicking the App menu, highlighting the Pictures sub-menu, and then clicking MATE Image Eye.

Image Viewer supports image file formats. The following image formats can be opened: ANI, BMP, GIF, ICO, JPEG or JPG, PCX, PNG, PNM, RAS, SVG, TGA, TIFF, WBMP, XBM, and XPM. Image Viewer supports the following storage formats: BMP, ICO, JPEG or JPG, and PNG. Image Viewer may open and save some image formats, depending on your system configuration and other software installed.

Elements of the Image Viewer

- **Menu bar:** The menu bar menus contain all the instructions you need to work with images in Image Viewer.

- **Toolbar:** It contains a subset of instructions that you can access in the menu bar.

- **Showcase:** The display area shows the image file.

- **Status bar:** The status bar provides information about the image. To show or hide the status bar, select View ▶ The status bar.

- **Photo Collection:** Image Gallery shows you all supported images in the currently active directory. It appears when the image has been uploaded. To show or to hide the collection, select View ▶ Photo Collection or press F9.

- **Image Collection window:** The Image Collection window provides thumbnails of images in the same folder as the image in the display area. Showing or hiding a file in the photo collection window, select View ▶ Image Collection.

MATE Calculator

MATE Calculator started out as a gnome-calc fork, a calculation system located in the OpenWindows Deskset of the Solaris 8 operating system.

MATE System Monitor

The mate-system-monitor allows you to monitor and control processes that run on your system. You can access detailed memory maps, send signals, and cut processes.

In addition, the mate-system-monitor provides a complete overview of application usage on your system, including memory and CPU allocation, as well as network usage. It also allows you to view file system information such as Device, Type, Mountpoints, and Disk usage. The System tab will display basic information about your system such as hostname, Kernel, MATE version, installed memory, and processing information.

MATE Terminal

You can use a deadly simulation app to access the UNIX shell instead of MATE. You can use any application designed to work on VT102, VT220, and xterm terminals with it. MATE Terminal also has the ability to use multiple terminals in a single window (tabs) and supports different configuration management (profiles). MATE Terminal is a fork of GNOME Terminal.

ADDITIONAL MATE PACKAGES

Several other illegal MATE applications are provided and maintained by the MATE community, so they are not included in the partner or partner group.

- Dock Applet

- Applet Online Radio

- MATE Menu

- MATE Desktop

- Brisk Menu

The first package of MATE is given below.

Dock Applet

The MATE Dock Applet is a MATE panel applet that displays open windows/applications as icons. The latest version of 0.78 includes five new indicators, a new option to add space between dock icons, and more.

Among the MATE Dock, Applet features are pinning apps in the dock, showing apps usage indicators, supporting updating apps using keyboard shortcuts, and more. The applet can even change the color of the MATE panels to prominent desktop background.

Changes to the MATE Dock Applet 0.78 include:

- Five new app indicators: circle, square, triangle, diamond, and subway. With the GTK3 version of the applet (Ubuntu 16.10 and later), the cursor color will be the current theme's highlight color. As that does not happen in GTK2, you can use the color back option to set this up (see the MATE Dock Applet preferences, on the Misc tab).

- Now, you can set the space between the dock icons. 0-7 Supported Values (Dock Preferences> Panel Options> Application Space).

- In windows that need attention, you can now adjust if the badge (exclamation mark) should be displayed at the top of the icon instead of the flashing icon (Dock Favorites> Mixed> Action when apps need attention).

Installation of Dock Applet

The MATE Dock Applet is available in Ubuntu storage (MATE), but it is not the latest version. You can see the version available for each Ubuntu release HERE. To install the version from the official Ubuntu MATE archives, use the following command:

```
$ sudo apt to install mate-dock-applet
```

Ubuntu MATE 17.04, 16.10, 16.04, or 14.04 users can install the latest MATE Dock Applet using WebUpd8 MATE PPA. Add PPA and install the applet using the following instructions:

```
Sudo add-apt-repository ppa: webupd8team / mate
$ sudo apt update
$ sudo apt to install mate-dock-applet
```

Once installed, right-click the MATE panel, select "Add to the panel," and add the "Dock" applet.

APPLET ONLINE RADIO

A panel applet lets you play your favorite online radio station on your system panel with a single click. Includes an extensive list of online radio stations, each with thousands of lists:

- Ice

- Radio browser

The applet should run on modern MATE Desktop and GNOME2 assets (e.g., RHEL 6 and all other derivatives like CentOS). On the MATE desktop, the applet supports building with GTK2 or GTK3. You must use the same version your MATE Desktop does.

Usage of Applets

When the applet starts, it loads the last radio station you listened to. Click on the applet to change its status from "Pause" to "Play"; click again to pause. The icon shows the state (playing or paused). Now, right-click the applet and go to the "All" menu to manage your channels.

- On the "Favorites" tab, you can add new channels (name and URL are required, URL includes a rule of thumb), remove existing ones, or start playing one. You can also rearrange the channels there. The list will be saved for each change.

- On the "Icecast" tab, you can download the complete Icecast directory (to be saved later), search through it, listen to the channel, or copy one to the favorites list.

- On the "Custom" tab you can enter your list of channels. It should only match the 4 most important fields of the Icecast XML format:

```
<directory>
 <entry>
  <server_name> Radio 1 (AAC +) </server_name>
  <listen_url> http://icecast.omroep.nl/radio1-bb-aac
                  </listen_url>
```

```
<type> Normal </type>
 <bitrate> 128 </bitrate>
</ login>
<entry>
 ...
</ login>
 ...
</litext>
```

When the applet launches, it will load with the latest ten channels and top ten favorites in two separate submenus in the right-click menu. From them, you can open the channel again. Note: Due to MATE/GNOME2 restrictions, these two submenus cannot be updated automatically and will only change at the beginning of the applet.

When a song starts playing, a notification will display. If you do not want to be interrupted every few minutes, use the Options tab in the All Channels menu to disable notifications.

MATE MENU

It enhanced the MATE menu. It supports filtering, preferences, easy release, auto session, and many other features.

This menu originated from the Linux Mint distribution and was distributed to other distros that sent the MATE Desktop Environment.

It is the MATE Menu, the MintMenu fork.

- The MATE menu removes specific Mint search options.

- The MATE menu removes package management features.

MATE DESKTOP

It now has a graphic design tool called "MATE Tweak," developed by Martin Wimpress, the founder and project leader of Ubuntu MATE. MATE Desktop is a continuation of the old GNOME 2. It provides an intuitive and attractive desktop environment using common Linux metaphors and other applications like UNIX. Ubuntu MATE is a community flavor of Ubuntu using MATE Desktop. It has released Ubuntu MATE 14.04 LTS and Ubuntu MATE 14.10. MATE Tweak is a simple configuration tool written in Python. It is a mint desktop fork that removes specific Mint configuration options.

MATE TWEAK

MATE Desktop now has a graphic design tool called "MATE Tweak," developed by Martin Wimpress. The founder and project leader of Ubuntu MATE. MATE Desktop is a continuation of the old GNOME 2. The developer has developed packages for his Ubuntu MATE PPA, which are available for testing at Ubuntu 15.04, Ubuntu 14.10, Ubuntu 14.04, and Linux Mint 17.

Installation

To add a PPA and enter a MATE Tweak, press Ctrl + Alt + T and then use the instructions one by one.

```
Sudo add-apt-repository ppa: ubuntu-mate-dev /
                trusty-mate
$ sudo apt-get update
$ sudo apt-get inserting mate-tweak
```

Brisk Menu is an open-source menu made for the desktop Mate area, which is usually delivered via Solus OS as the default menu applet. However, Brisk has its functionality, e.g., a built-in search feature that mimics the original Windows menu while still providing high performance.

It has a flexible, portable UI and compresses your battery and pleasant memory. It does not surprise me, especially after learning that Brisk-menu is a collaborative activity between Solus and Ubuntu MATE.

BRISK MENU

Brisk Menu is an open-source menu made for the desktop Mate area, which is usually delivered via Solus OS as the default menu applet. However, Brisk has its functionality, e.g., a built-in search feature that mimics the original Windows menu while still providing high performance. It has a flexible, portable UI and compresses your battery and pleasant memory.

Features

- Freeware: Brisk Menu is free for anyone who uses Mate Desktop to download and use it

- The customizable UI is fully customizable and intuitive

- List of favorites

- Fast performance

- Hotkey menu action support

- Support for desktop actions, e.g., context menus for desktop actions such as open incognito mode

- Sort the application and list of files by categories

- Session controls/screen saver

- Drag and drop support when working with launchers

- Sidebar launcher support

- Beauty options are available as GTK3 + CSS style options

- The Brisk Menu developers plan to bring additional changes in the future, including updating the UI Settings to control other visual features (labels/icons/options) and improving the style of a few window sections

Add the following PPA and install Brisk Menu.

```
$ sudo apt-add-repository ppa:flexiondotorg/brisk-menu
$ sudo apt update
$ sudo apt install mate-applet-brisk-menu
```

Select MATE from the menu in the display manager of your choice. The display manager, or login manager, is usually a visual application displayed at the end of the startup process instead of the default shell. There are different implementations for display managers, just as there are different types of window managers and desktop areas. There is a certain amount of customization and content available for each. Or, to start MATE with startx, enter exec mate-session in your ~ / .xinitrc file. See xinitrc for details, such as keeping login time.

INSTALLATION ON VARIOUS OS

- **Fedora Linux:** The applet is now available in the RPFusion Free Fedora repo

  ```
  yum install mate-applet-streamer
  ```

- **Arch Linux:** Applet is available in Arch's repo

- **Mageia Linux:** Applet is available in Mageia repo

- **RHEL6 and other alternatives (CentOS etc.) that use GNOME2:** RPMs provided below (or applet can be installed in the source).

DISPLAY MANAGERS

There are various display managers used in MATE Desktop Environment as given below:

1. CDM

2. Nodm

3. Console TDM

4. Ly – TUI display manager

5. Tbsm – Station-Based Session Manager

You will get brief introduction of each display manager.

CDM

So basically it is a minimalistic but full-featured feature instead of a portable controller such as SLiM, SDDM, and GDM that provides a fast-paced entry, chat-based X Window System header. Written with pure bash, CDM is virtually non-dependent yet supports many users/sessions and can launch almost any desktop or window manager.

Console TDM

The TDM display manager is a startx text. The main branch of development is development. The main branch is used for extraction. It has an assistant text called tdmctl to manage tdm times. Each tdm session is a usable script connector that will be called instead of the last exec of .xinitrc.

The following are three types of tdm sessions.

1. **X:** This session started within the .xinitrc file as part of a startx call (apply this to the X window manager/desktop areas).

2. **more:** this session starts in the shell (apply this to Wayland sessions, tmux wrappers, etc.).

3. **Tdm session may or may not work:** An active session is present in the tdm selection screen, while an inactive tdm session is not displayed. If the path is not a usable file, the tdm session is always invalid.

Installation

Execute "make install" from the source directory (you can choose to set DESTDIR or PREFIX). There are following dependencies:

- xinit

- dialog box (optional, cursing interaction)

- **Usage:** To add tdm to a local user, use

```
tdmctl init
tdmctl add <time name> <usable mode> [X (default)
                / additional]
```

- It will copy the tdm configuration directory to your home directory. You must then edit your .profile file (or .bash_profile, .zprofile, etc.) in order to call tdm as the last command (this will launch tdm once you log in to tty). If you want to allow multiple X times, you should use the option --disable-xrunning-check. In your .xinitrc file, you must change the exec line with exec tdm --xstart, which will start your X period (if you do not have an .xinitrc file, create a new one with this line).

- **Return value:** If no session is used, return 1. If not 0.

Texts

There are two texts in the TDM configuration guide ($ HOME / .tdm) used at the beginning and end of tdm.

1. Tdminit is run in front of the selection screen (when tdm is called out --xstart)

2. tdmexit was created before the startx command was called (in tdm --xstart run)

tdmctl Commands

- Launch the configuration directory.

  tdmctl init

- List of available (active) sessions.

  Tdmctl list: list available sessions (X and more)

- List of cached (inactive) sessions

    tdmctl repository

- See which command is called the session

    tdmctl check <shop>

    tdmctl check out more/<session>

- Show or set default session

    default tdmctl [session]

- Add session (fast)

    tdmctl add <name> <mode> [X (default) / extra]

- Delete session

    tdmctl delete <session>

- Enable or close the session.

    tdmctl enable / disable <page>

- Move the configuration to the compliant XDG directory

    tdmctl move

Nodm

It is a small display manager that logs in as a given user and starts an X session without asking for a username or password. Using nodm is a major security issue on a normal computer because it can give anyone access to a computer. However, there are various cases where automatic login is required: for example, in an embedded system such as a cell phone, a kiosk setting, or the control panel of industrial equipment. In those cases, nodm is easy to set up, lightweight, and should do the right thing.

Features

nodm is as small as it can be, and it tries to provide the minimum number of features required to do a good job, very much following the goal of a little surprise. Here is what is offered:

- Automatic login with a rooted user, doing all that needs to be done such as setting up a session with PAM, updating lastlog, logging in to syslog.

- Nodm makes the VT offer, looks at the free terminal where it will use the X, and keeps it redistributed to X restart.

- X launched (by default, / usr / bin / X).

- When X esrver is ready to accept the connection, an X session is set:

  - The DISPLAY and WINDOWPATH variables are set.

- The session wrapped in a PAM session, which sets the user's location

- ~ / .xsession-error is terminated if any.

- Session text is run (automatically, / etc / X11 / Xsession) using "sh -l."

- If it is an X server or X session exit, one is killed, and both are restarted.

- If the session is out soon, nodm will wait a bit before restarting. The waiting times are as follows:

  - The first time a session is out soon, restart immediately.

  - Second and third, wait 30 seconds.

  - For the rest of the time, wait 1 minute. If the session lasts long enough, the waiting time returns to zero.

- Nodm is NOT COMING yet and works in the background as a proper daemon: more distribution than tools do that, and nodm plays well with it. It is not a design choice: simply, so far no one has ever felt the need to use it.

Configuration

The configuration is done with the following variables:

- NODM_USER: Controls the user used for automatic login.

- NODM_X_OPTIONS: X server command line (example: "vt7 -nolisten tcp"). Extensions using wordexp, with tilde extensions, dynamic changes, arithmetic extensions, wildcard extensions, and quote deletions, but no replacement command. If a change of order is required, please contact and provide the actual.

- NODM_MIN_SESSION_TIME: The minimum duration (seconds) for a session to last for nodm to determine has not stopped yet. If session X continues for less than this time, nodm will wait for a growing amount of time before restarting (default: 60).

- NODM_XSESSION: X session command (default: / etc / X11 / Xsession). It works using a shell so that it can be any shell command.

- NODM_XINIT was Used for older versions of nodm as the path to the xinit program, but is now overlooked.

- NODM_X_TIMEOUT: Closing time (seconds) to wait for X to be ready to accept the connection. If the X is not ready before this closing time, it is executed and restarted.

Ly – TUI Display Manager

Ly is a lightweight (such as ncurses) TUI manager for Linux and BSD. There are various dependencies, as given below.

- C99 compiler (tested with tcc and gcc)

- standard C library

- GNU does

- pam

- xcb

- xorg

- xorg-xauth

- mcookie

- tput

- shut down

In Debian-based distros using apt install build-essential libpam0g-dev libxcb-xkb-dev as root should add all that depends on you. The following desktop areas were successfully supported Ly display:

1. budgie

2. cinnamon

3. deepin

4. enlightenment

5. gnome

6. i3

7. kde

8. lxde

9. lxqt

10. partner

11. move

12. xfce

13. the pantheon

14. maxx

15. window maker

Ly should work with any X desktop space, and provide basic wayland support. The supports are given below:

Consolidation and Integration

- Clone the storage repository

```
$ git clone --recurse-submodules https://github.com/
    nullgemm/ly.git
```

- Combine

$ make

- Check-in default tty (automatic tty2) or terminal emulator (but desktop situations will not start)

```
$ sudo make run
```

- Install Ly and the systemd serviced file provided

```
$ sudo make install
```

- Enable the service

```
$ sudo systemctl enables ly.service
```

- If you switch between ttys after Ly's start, you should also disable getty in Ly's tty to prevent "login" from reproducing over it

```
$ sudo systemctl disable getty@tty2.service
```

Controls

You can use the keyboard up and down keys to change the current field the left and right arrow keys to change the target desktop while in the desktop field.

Tbsm – Station-Based Session Manager

It is a clean bash session or app launcher, inspired by cdm, tdm, and krunner. Tbsm is a program or session launcher, written in pure bash without ncurses or chat dependencies. It is inspired by cdm, tdm, and somehow by krunner. Attempts were made to design the tbsm behavior as simple as possible and start daily activities with the main lower lashes.

- **Support:**

  - Tbsm has a low operating ecosystem.

  - It has 94 stars with 11 forks.

  - It hasn't had a major release in the last 12 months.

  - On average, the problems are closed within 16 days.

  - It has a neutral feel to the engineering community.

- **Quality:** Tbsm has no reported problems.

- **Security:** Tbsm has no reported risks, and libraries relying on it have no reported risks.

- **License:** Tbsm is licensed. The patents licenses can be open source licenses that do not comply with the SPDX, or licenses are not open source, and you need to update them closely before using them.

- **Reuse:** tbsm release is not available. You will build from the source code and install.

GRAPHICAL DISPLAY MANAGER

There are various graphical display managers used in MATE Desktop Environment.

1. Entrance-Based EFL display manager

2. Enlightenment display manager

3. GNOME Display Manager

4. GDM Face Browser

5. LightDM

6. SDDM

7. Xdm

There are various display managers used in MATE Desktop Environment as given below.

Entrance-Based EFL Display Manager

It is a fork and a current version of development. IT'S ALIVE! IT WORKS – login to Unix Display/Login Manager listed in Enlightenment Foundation Libraries (EFL). Login allows users to select an X WM/Desktop session to launch it when they have successfully logged in. The entry is live and active again by logging into the X sessions and finally at the Wayland session!

How to Use It?

To get started login, you need an init system script or systemd (unchecked). It may vary based on your operating system. The login does not provide an init script at this time, and it may not work or function properly if it is started directly. Login should be called init script or systemd service. There is a system log file provided for login. It is not known whether this works or not.

Login User

Login now starts as a root and then uses the setuid to run the entrance_client under the non-user. It is done with sudo and su for a while before switching to setuid. The login was also used to operate under the "login" user but was changed to a previous commitment. Inherited design is from second generation. It may be modified so that the login is operated under its user interface and is not configured or operated under root. It will require creating a user account, adding user video permissions, etc. It is possible to accomplish this now, start logging under user "login" by adjusting the unchecked login option.conf start_user: value "start_user" string: "entrance"; You will also need to create an access directory and ensure it has the appropriate permissions. Since login will not work under root, login will not fix this, despite having a code for that. If started as a root, login will create the proper permissions as needed.

This includes the Lighting window manager and the Basic Lighting Library (EFL) libraries, which provide additional desktop environment features such as a tool kit, object canvas, and hidden objects. It has been under development since 2005. But in February 2011, the main EFLs saw their first stable release of 1.0. Development PKGBUILDs are downloading and installing the latest development code available as enlightenment-gitAUR and its dependents.

The following are EFL-based applications, many of which are in the early stages of development and have not yet been released:

- **Ecrire-gitAUR:** Ecrire text editor

- **ediAUR:** EFL-based IDE

- **eliminance-gitAUR:** Eluminance Image Browser

- **enjoy-gitAUR:** Enjoy the music player

- **EperiodiqueAUR:** Eperiodique periodic table viewer

- **Ephoto and ephoto-gitAUR:** Ephoto Photo Viewer

- **epourAUR:** EFL-based Torrent client

- **epymc-gitAUR:** E Python Media Center

- **equate-gitAUR:** Rate the calculator

- **eruler-gitAUR:** Eruler on-screen Eruler and measurement tools

- **efbb-gitAUR:** Escape from the angry bird style game in Booty Bay

- **elemines-gitAUR:** Elemines minesweeper style game

- **rage and rage-gitAUR:** Rage video player

- **terminology-gitAUR:** Git master current terminology

Installation

You may also want to enter words, an EFL-based terminal emulator that integrates well with Enlightenment. If the article invites you to install other packages in the usual way, it will not show detailed instructions. Instead, it will simply state the names of the packages to be installed. Note: Generally, embedded or embedded links are used to identify this article section. However, JavaScript must be turned on for these links to work. The sub-sections summarize the standard installation procedures depending on the package type.

Themes

Additional themes for customizing the look of the Enlightenment are available:

- enlightenment-themes.org

- relighted.c0n.de of default themes in 200 different colors

- git.enlightenment.org (git clone theme you like, use "mark" and end with a .edj theme file)

- packages.bodhilinux.com has a great collection (you will need to download the .edj file to .deb; bsdtar will do this, and it is part of the installation of Arch Linux). A good catalog can be seen in their wiki

- exchange.enlightenment.org (archive)

- enlightenment-themes.org

- relighted.c0n.de of default themes in 200 different colors

- git.enlightenment.org (git clone theme you like, use "mark" and end with a .edj theme file)

- packages.bodhilinux.com has a great collection (you will need to download the .edj file to .deb; bsdtar will do this, and it is part of the installation of Arch Linux). A good catalog can be seen in their wiki

- exchange.enlightenment.org

GNOME Display Manager

A display manager uses all the essential features needed to manage attached and remote displays. It was written from scratch and doesn't contain XDM code or X Consortium. Note that GDM is configurable, and many configuration settings affect security.

GDM 2.20 and pre-established stable workstations. However, the codebase was completely rewritten in GDM 2.22 and is not entirely compatible with older releases. It is partly because things work differently, so some options are irrational, partly because some options have never made sense, and partly because some functions have not been restarted. Visible connectors that continue to receive consistent support include Init, PreSession, PostSession, PostLogin, and Xsession scripts. Some configuration options in the <etc> /gdm/custom.conf file continue to be supported. Also, ~ / .dmrc, and browser face image areas are still supported.

GDM 2.20 and supported handling multiple displays with different image cards, such as those used in terminal server locations, enter through windows like Xnest or Xephyr, gdmsetup program, XML-based greeting themes, and the ability to use the selector XDMCP on login screen. These features were not added during the 2.22 rewrite.

Performance

GDM is responsible for managing the indicators in the system. It includes authenticating users, starting user time, and ending user sessions. The GDM is adjustable, and the configuration options are described in this document's "GDM Configuration" section. GDM is also accessible to users with disabilities. GDM provides the ability to handle a large console display with VT-enabled displays. It is integrated with the Fast User Switch Applet (FUSA) and GNOME-screensaver to take multiple shows in the console via the Xserver Virtual Terminal (VT) interface. It can also manage XDMCP displays. GDM will do the following when managing the display regardless of the type of display. It will start the Xserver process, use the Init script as the root user, and start the greeting program on display.

GDM and PAM can be configured not to require any input, which will allow GDM to log in automatically and start the session, which can be helpful in some areas, such as single-user systems or cookies. In addition to confirmation, the greeting system allows the user to choose which session to start with and which language to use. The descriptive sessions are files ending in the .desktop appendix. By default, it is configured to display a face browser to select their user account by clicking the image instead of typing in their username. GDM tracks the user's default time and language in the users ~ / .dmrc and will apply this default if the user did not select a session or language in the login GUI.

After verifying the user, the daemon uses the PostLogin script as root and launches the PreSession script. After using these scripts, user time is started. When users exit their session, the PostSession script is used as root. These documents are provided as distribution hooks and end-users to customize how times are handled. For example, using these hooks, you can quickly set a machine that creates a $ HOME user directory and delete it from the exit. The difference between PostLogin and PreSession texts is that PostLogin starts before the pam_open_session call, so it is an excellent place to do whatever needs to be done before the user session is started. The PreSession script is called after the start of the session.

Accessibility

GDM supports "Accessible Login," which allows users to access their desktop session even if they can't use the screen, mouse, or keyboard easily in the usual way. Affordable technology (AT) features such as an on-screen keyboard, screen reader, screen magnifier, and Xserver AccessX keyboard access are available. If needed, it is also possible to enable large text or high-resolution icons and controls. See the "Accessibility Configuration" section of the document for more information on how various accessibility features can be configured.

GDM Face Browser

Face Browser is a visual interface that allows users to select their username by clicking on the image. This feature can be enabled or disabled by the key org. gnome. login-screen disable-user-list GSettings and opens automatically. If disabled, users must type their full username manually. When enabled, it shows all available local users to sign in to the system (all user accounts defined in the file/etc./passwd with valid shell and UID high enough) and remote users recently logged in. The face browser in GDM 2.20 and earlier will show all remote users, which has caused performance problems in large business installations. Face Browser is designed to display users who regularly log in to the top of the list. It helps ensure that regular users can quickly access their login images.

Face Browser supports "Previous type search," which moves face selection as the user types in the corresponding username in the list. It means that a user with a long username will only need to type in the first few letters of the username before selecting the appropriate item in the list.

GDM will use the "stock_person" icon defined in the current GTK + theme if the user does not have a defined face image. If no such image is defined, we will return to the normal face image. Please note that uploading and measuring face icons found in a remote user's home directory can be time-consuming. Since uploading images via NIS or NFS does not make sense, GDM does not attempt to upload face images to remote home lists.

When the Browser is turned on, computer-enabled usernames are displayed for everyone to see. When XDMCP is enabled, usernames are displayed to remote users. It, of course, somehow limits security as the malicious user does not need to guess the valid usernames. In some restricted areas the face browser may not be suitable.

LightDM

It is a free X display manager and open source that aims to be easy, fast, expandable, and multi-desktop. It can use different ends to draw User Interfaces, also called Greeters. It also supports Wayland.

It is the default display manager for Edubuntu, Xubuntu, and Mythbuntu from 11.10 released Lubuntu from 12.04 released to 16.10, and Kubuntu from 12.10 to 15.04 for Linux Mint and Antergos.

The following features are included:

- It is a codebase with very few dependencies

- It supports different display technologies (X11 and Wayland using Mir)

- It supports remote login (incoming – XDMCP, VNC, outgoing – XDMCP, connected)

- Comprehensive test suite

- Compliance standards (PAM, login, etc.)

- A good interface between the user and the server interface

- Cross-desktop (greeters can be written in any tool kit)

- A well-defined greeting API that allows multiple GUIs

- Support for all conditions of use of manager-display, and plug-in where appropriate

- LightDM has a more straightforward code base than GDM and does not load any GNOME libraries to work.

SDDM

It is a modern X11 and Wayland display manager that aims to speed up, simplify, and improve. It uses state-of-the-art technology such as QtQuick, which gives the designer the ability to create smooth, vibrant user links.

SDDM has a great theme. We do not set limits on the design of the user interface, and it depends entirely on the designer. We provide a few callbacks to the interface that can be used to authenticate, pause, etc. We provide some pre-made features like text boxes, combox, etc.

XDM

It contains a collection of X indicators, which may be on a local host or remote servers. The xdm design was guided by the requirements of the X terminals and the standard XDMCP of the Open Group, the X Display Manager Control Protocol. It provides services similar to init, getty, and login to character terminals: password entry and password, user authentication, and use "session." XDM provides specific login information.

X DISPLAY MANAGER PROTOCOL

The X Display Manager Control Protocol (XDMCP) uses the UDP 177 port. The X server requests the display manager to start the session by sending a query package. If the display manager allows access to that X server, it responds by sending the favorable package back to the X server. The display manager should verify himself on the server. To do this, the X server sends the application package to the display manager, which returns the Receive package. The display manager is authorized if the Accept package contains the X server response. For example, generating the correct response may require the display manager to have access to a private key. The X server sends a Manage package to notify the display manager if verification is successful. The display manager then displays its login screen by connecting to the X server as a normal X client. During the session, the server may send KeepAlive packets to the display manager from time to time. Suppose the display manager fails to respond to the Alive packet within a certain time. In that case, the X server assumes that the display manager has stopped working and may terminate the connection.

Configuration

MATE can be configured through its Mate-control-center program, offered in a mate-control-center package. To control some computer hardware, you may need to install additional tools.

- Backends that are supported by the mate-media package are ALSA and PulseAudio.

- For Bluetooth device support, you need to install the blueman package.

- To configure the network, install the network manager applet package. See NetworkManager.

- The mate-power-manager package supports the Power backend.

- To configure printers, install the system-config-printer package.

Accessibility

MATE is well suited for use by those with limited vision or mobility. Install orca, espeakAUR (Screen reader for the blind or visually impaired), and onboard (On-screen keyboard is useful for users who can walk). Before starting MATE, enter the following command as a user requiring accessibility features:

```
$ gsettings set org.mate.interface accessibility is
            real
```

Once you have started MATE, you can set up accessibility applications using System> Preferences> Help Technology. Although you need Orca, you will need to run it from the Alt-F2 window to start getting the conversation.

Notifications

- **Battery discharge:** To disable notification on battery discharge, use:

```
$ gsettings set org.mate.power-manager notify-
            executes false
```

Ubuntu MATE Workstations

- Workstations using Ubuntu MATE 16.04.

- The latest LTS release for Ubuntu OS.

- Ubuntu MATE is derived from Ubuntu and upgraded for easy installation and use.

- Internally, Linux is robust and secure without the need for antivirus protection.

- Ubuntu MATE is very portable and incorporates 3D graphics into laboratories.

- However, it will also apply to slightly robust graphics processors.

- MATE includes user-based menu interactions where users can look at the installed programs mentioned in the sections.

- It makes installed applications relatively easy to search.

- MATE is one of the prominent Linux software executives.

- We can use the software administrator to find the app if we do not find it.

DISTROS SUPPORTS MATE

MATE exists through the official Linux distros repository mentioned below:

- Void Linux

- Vector Linux

- Uruk Linux/GNU

- Ubuntu MATE

- Ubuntu

- Trisquel Linux/GNU

- Solus

- Slint

- Salix

- Sabayon

- PLD Linux

- PCLinuxOS

- Parrot Security OS

- openSUSE

- Mananjaro

- Mageia
- Linux Mint
- Hamara Linux
- GNU GuixSD and GNU Guix
- Gentoo
- Fedora
- Debian
- Arch Linux
- AOSC
- Antergos
- Alpine Linux

MATE DESKTOP APPLICATIONS

- Engrampa
- Caja
- Pluma
- Calculator
- Font Viewer
- Search
- Screenshot
- Colour Selection
- Dictionary
- System Log
- Disk Usage Analyzer
- System Monitor
- Terminal
- Power Statistics

- Mozo
- Déjà Dup
- GNOME Disks
- Character Map
- Passwords and Keys
- Plank
- Redshift
- Firefox
- Simple Scan
- Shotwell
- Transmission
- Evolution
- LibreOffice
- Celluloid
- Cheese
- Webcamoid
- Rhythmbox
- GParted
- gdebi
- Firmware
- Blueman
- Network Manager
- gufw
- Magnus
- Onboard
- Orca Screen Reader

HIGHLIGHTS

MATE is a fork of GNOME 2, one of the most popular Linux desktops. MATE follows standard desktop paradigms and provides you with a visually pleasing and intuitive interface. It is incredibly sharp out of the box with a friendly theme and icon pack. The feel is good, too, using a minimal app and a fast visual interface. Things are going well, and MATE can be a great choice for a low-level system or for those who want to stay efficient in system resources.

- **User Experience:** MATE feels like a unique mix of the old, traditional desktop with a different Linux twist. For older MATE applications, there is a menu at the top left with a list of categories for all your applications, as well as a search function. At the top right is a well-designed system tray with notifications, network, audio, and time is at the top right. At the bottom left is a "Show Desktop" button that helps many heavyweights or those who like to use desktops and desktop icons. There is also a trash can in the bottom right. All you will need is the menu at the top left. A robust menu that looks at application names and application descriptions.

- **Caja File Manager:** It is the most important application on the MATE desktop. It is a good file manager with a lot of work. Out of the box, it supports several different side windows, including a stock list menu, tree view, directory history, and document information. Additionally, there is a simple reset button: /// locale to see all available disks, drives, and available file systems.

- **Pluma Text Editor:** Pluma Text Editor is an excellent Gedit fork that adds some friendliness to users. For one thing, all the options for saving, opening new files, and finding or retrieving and replacing are all in the top bar rather than hidden behind the hamburger menu. If not, Pluma is a straightforward and easy-to-use text editor that gets in your way and is easy to use.

- **Search Tool:** The MATE Search Tool is a great way to look at all your files in your system. Similar to Catfish in Xfce, but there are more options available. You can search files by name, content, modified date, owner, group, size, and similar word expression patterns. It's a great way to work with files in your system.

- **Customization:** MATE may not be as customized as KDE Plasma, but you can change it slightly. You can change the menu icons, panel layout, themes, look, and add and remove panels and docs. One great thing about MATE is that you can select and select different pieces of different themes and combine them to create a custom theme of what you want. Additionally, you can customize the MATE workflow add new workspaces. All in all, MATE happens to be customizable logically.

- **Performance:** MATE has excellent performance. The new Ubuntu MATE 20.04 boot uses 478 MB of RAM and approximately 1% CPU to keep things running. That's a lot less, especially when considering the fully featured desktop you use. MATE will work best on older systems with less RAM and anemic CPUs. MATE feels fast to use. Applications open quickly, workplaces change without delay, and windows open easily when tiling the sides and corners. It closely follows traditional desktop paradigms, and it sounds a good fit for an older system that requires an extra retro vibe or feels comfortable in the old paradigm.

CHAPTER SUMMARY

In this chapter, we have covered the introduction of MATE: its installation, features, history, core projects, applications, and development. Also, we have a separate section where you get a history of MATE versions.

CHAPTER 6

Budgie Desktop Environment

IN THIS CHAPTER

> ➤ Introduction

> ➤ History

> ➤ Budgie versions

> ➤ Installation

> ➤ New features and development of Budgie

In the previous chapter, we learned about MATE. In this chapter, we will briefly discuss one of the flavors of Budgie.

INTRODUCTION

It is a desktop site currently using GNOME technology like GTK (> 3.x). Participants are developing the Solus project from many communities such as Arch Linux, Manjaro, openSUSE Tumbleweed, and Ubuntu Budgie. Budgie's design emphasizes simplicity, minimalism, and elegance. The Solus project will replace the GTK library with the Enlightenment Foundation Library (EFL) to release Budgie 11.

Budgie's current desktop integrates seamlessly with the GNOME stack, using basic technology to provide desktop information. Budgie applications typically use GTK and headers similar to GNOME applications.

DOI: 10.1201/9781003308676-6

Budgie automatically creates a list of favorites as the user works, moving categories and apps pop-up in the menus when in use.

WHAT IS UBUNTU BUDGIE?

Ubuntu Budgie is an advanced community distribution, including the Budgie Desktop Environment and Ubuntu at its core. Whether you use it on a computer or a powerful workspace, Ubuntu Budgie is a flexible operating system on any device, which keeps it fast and usable.

It includes a well-tested and stable Ubuntu Core and a traditional, lightweight, and modern desktop integrated with the Solus project. The version of Ubuntu Budgie is 21.10, released on October 14, 2021.

HISTORY OF UBUNTU BUDGIE

Budgie was initially developed as an automated desktop distribution platform for Evolve OS Linux. After Evolve OS was changed to Solus, Budgie's development accelerated. Budgie's earlier versions were slow and often fragmented. Speed and reliability are improved with successive releases. Budgie v1 was released on February 18, 2014, and v10 on December 27, 2015. The version has changed, with the current release of 10.5.

On September 14, 2021, the Solus project announced that the upcoming release of Budgie 11 would no longer be recorded on GTK due to unresolved disagreements with the GNOME team. The GNOME-automated software will also be replaced in the next Solgie Edition of Solus.

The Ubuntu Budgie version started as a taste for the illegal community in line with the Ubuntu 16.04 LTS release known as the budgie-remix. The 16.10 version of the budgie-remix was later published following strictly a timeline which was a problem for the 16.10 version of Ubuntu.

Eventually, it was identified as the taste of the official Ubuntu community and renamed Ubuntu Budgie. 17.04 version of Ubuntu Budgie was published in April 2017 and upgraded to version 17.10 in October 2017. In the 18.10 version of Ubuntu Budgie, 32-bit support has been reduced. 32-bit support has been reduced to Ubuntu MATE flavor as well. Vincenzo Bovino worked as a PR and a new product manager.

THE RELEASE OF THE UBUNTU BUDGIE

There are various releases of Ubuntu Budgie. Let's discuss each in brief.

1. Budgie-remix 16.04

2. Budgie-remix 16.10

3. Ubuntu Budgie 17.04

4. Ubuntu Budgie 17.10

5. Ubuntu Budgie 18.04

Budgie-Remix 16.04

The 16.04 version of the budgie-remix was released on April 25, 2016, four days after the launch of the 16.04 version of Ubuntu. The program includes version 10.2.5 version of Budgie Desktop. It has a Mutter-based window manager, notification center settings, and a customizable panel.

The theme, Arc-GTK +, is used in this release. Budgie-remix has a 3.18 version of Nautilus, as at the time Ubuntu OS had a 3.14 version of Nautilus compared to version 16.04 of Ubuntu.

It includes 0.11.1 version of dock panel Plank, version 3.18 gedit, version 3.18 of GNOME images, version 3.4.3 of gThumb, version 3.3 of Rhythmbox, version -3.18 for Totem, version 3.18 of GNOME Terminal, version 5.1.2 version of LibreOffice, version 45.0 of Mozilla Firefox, version 2.84 of Transmission, version 3.18.9 of -GTK +, version 11.2.0 for Mesa, version 1.18.3 for XOrg, and version 4.4.0 version for Linux kernel.

Budgie-remix is an illegal Ubuntu flavor; however, its developers aim to make it an official member of the Ubuntu family. The goal is to release Budgie-remix 16.04.1 in three months, followed by the first alpha of Budgie-remix 16.10 in July. It is based on Ubuntu 16.04 LTS and includes the latest Budgie Desktop 10.2.5. Budgie Desktop integrates seamlessly with the GNOME stack. It consists of a libmutter-based window manager and a custom panel providing an applet, notification, and a custom center, called Raven.

The new illegal Ubuntu flavor already combines its art with the custom theme of Plymouth and LightDM and the beautiful themes of Arc GTK and Faba. Since Budgie is the GNOME shell, Budgie-remix 16.04 delivers GNOME Settings (Control Center/Daemon Settings). In addition, Budgie-remix 16.04 includes Nautilus 3.18, not the version found in the official Ubuntu 16.04, which is 3.14.

That may be because Budgie-remix is not the official taste of Ubuntu, so it uses PPA automatically, not only providing Nautilus 3.18 but also the entire Budgie Desktop, artwork, and more. You can use the PPA to try Budgie Desktop on Ubuntu, but remember that it will update Nautilus to a version without Unity patches (you can lock the current version of Nautilus to avoid that).

Other apps automatically installed with Budgie-remix 16.04 include the Plank dock (0.11.1), which is used on the left screen with Intellihide and a nice theme called Arc automatically, and:

- Gedit 3.18

- Images of GNOME 3.18

- gThumb 3.4.3

- Rhythmbox 3.3

- Totem 3.18

- GNOME Eye (Image Viewer) 3.18

- GNOME Terminal 3.18

- LibreOffice 5.1.2

- Firefox 45

- Transfers 2.84

- And various resources such as Calculator, Disks, Calendar, and more.

Budgie-Remix 16.10

The budgie-remix 16.10 was released on October 16, 2016, three days after Ubuntu 16.10. The program contains the Budgie Desktop 10.2.7, GTK + 3.22, Linux kernel 4.8. Many new features are used in this version, such as full disk encryption, home folder encryption, and multilingual support during installation. It has Arc GTK + design theme icons for the new Pocillo theme.

What's New in Budgie-remix 16-10?

- Installation in any language.

- Full disk support and encryption for a personal folder.

- It includes the latest upgrades for budgie-desktop v10.2.7, including updates from Solus.

- Linux kernel 4.8.x.

- GNOME GTK + Applications 3.22.

- 16.10 is the background of the wallpaper contest.

- New budgie-welcome window.

- Option to switch from Arc team to Material Design.

- The arrival of your new Pocillo icons.

- Desktop applications updated.

To install new software and updates, Budgie-remix uses GNOME Software and Software Updater, such as Ubuntu.

Ubuntu Budgie 17.04

17.04 version of Ubuntu Budgie was released on April 19, 2017. Budgie-remix was renamed after the sharing became official in the Ubuntu community.

The program includes version 10.2.9 of Budgie Desktop, version 4.10 for Linux kernel, version 17.0.3 for Mesa, and version 1.19.3 for Xorg. The app, i.e., Budgie-Welcome, was updated, and system index support appeared. All audio apples were managed, 3.24 GNOME applications were activated, GNOME Terminal was replaced by Termix, and Google Chrome was replaced by -Chromium and support. The GTK + Qt theme is activated. Works with Arc GTK + design theme and Moka icons theme.

What's New in Ubuntu Budgie 17.04?

- Comes with Budgie Desktop Environment and Plank.

- It is not intended for mobile/tablet (like Unity/GNOME) but is for desktop use only.

- It consumes about 850 MB of memory by doing nothing.

- Comes with GNOME Software 3.22 and Calendar 3.24.

- Supports audio/video playback with MPV and Rhythmbox player.

- Provides easy-to-use customization with Raven and Menu Editor.

Built-in Software

Ubuntu Budgie delivers standard applications such as the standard Ubuntu, with some differences:

- LibreOffice (Author, Calc, Impress)

- Chromium Browser

- GNOME Apps: Calendar, Photos, Books, Weather, Maps

- Geary Mail

- System settings

- Rhythmbox

- MPV

- Terminix

System Details

Here is the list of some technical details about Ubuntu Budgie 17.04.

- Kernel version: 4.10

- LibreOffice 5.3.1

- Snap 2.23.6

- Terminix 1.4.2

- GNU bash version: 4.4.5

- Budgie 10.2.9

- Nautilus 3.20.4

- Chromium 56

- Plank 0.11.3

- GNOME Software 3.22.7 (it supports Flatpak and Snappy)

- GNOME Calendar 3.24 (it helps week view now)

Ubuntu Budgie 17.10

Version 17.10 of Ubuntu Budgie was released on October 19, 2017. The program includes version 10.4 of the Budgie Desktop shell and version 4.3

of the Linux kernel. Budgie's Desktop Environment 10.4 featured a group of native objects containing Alt + Tab with a new pattern, which supports changing window controls on the left or right side, Spotify support within Raven, encryption support. volumes, and SMB files are from applet extensions, and support customization occurrence from time to time. In addition, there are support for right or left side panels in version 10.4 of the Budgie Desktop, support for dashboard conversion, automatic dynamic and transparent image for all panels, and floor filling ability. Panel to be used without the effect "skip."

Other improvements to the edition of Ubuntu Budgie's 17.10 version include Caffeine and Night Light tools, Tilix as the ultimate (automatic) emulator than Termix, and Tilix Quake mode support F12 key. Also, it has new wallpapers and updated panel icons.

Canonical withdrew the 17.10 Ubuntu distribution in October and hid the link to download on December 20, 2017. The reason was a common distraction with BIOS damage on a few Lenovo notebook models and one Acer model.

Version 17.10.1 was published on January 12, 2018, which covers the resolution of the actual issue is version 17.10.

Ubuntu Budgie 18.04

The 18.04 version of Ubuntu Budgie was released on April 26, 2018. The version contains version 4.15 of the Linux kernel. It was possible to install OpenVNC on the network administrator.

Many new applets like DropBy have downloaded data on USB devices, Quick Note for all notes, Hot Corners to move windows in the corner of the screen, Window Mover to quickly remove windows from other visible desktops, automatic change keyboard layout in the program app, Clock functions to view time zones, Preview windows to view open windows, etc. The GNOME version has been updated to version 3.28, but Nautilus has version 3.26.

New Features and Enhancements

- Files (Nautilus) 3.26 have a Folder Color to change the color and add symbols to folders

- Files now come with Various Documents Templates available for use

- Settings (gnome-control-center) have a style change

- GNOME-based apps are now available in the latest version, 3.26 (if released)

- Budgie-welcome now has relevant title bars to clear the window

- The Budgie-welcome translation is now available in many languages

- One-click switch between IBUS and FCITX

- GNOME images have been replaced by gThumb

- The default Budgie reception settings now have stylish and extended edits

- Budgie can now incorporate many budgie applets from foreign companies

- The Global Menu Applet can be added to the panel

- Budgie installed snap links and flatpak websites for more software installation options

- The one-click theme has the best font support for key-based themes – fonts with better size

- Night-light and Caffeine (disable screen) are now sent as default panel icons

- Screen lock had a style change

- Move from Terminix to Tilix for a terminal template

- Support Tilix Quake mode – just press F12

- Ubuntu Background 17.10 Community Contest

- Added design wallpaper – used when using budgie-welcome makeovers

- Enhanced Panel Icons – stylish support for standard apps like Caffeine and Dropbox

- Login screen changed from lightdm-gtk-greeter to slick-greeter

The LTS version is supported for three years, while the standard release is supported for nine months. The new release includes a variety of fixes and improvements to the Ubuntu Budgie team that released it since the February 18.04.2 release.

The kernel and images found in version 19.04 have been restored in version 18.04. It is part of the Hardware-Enablement-Stack release, and 18.04 users, including HWE (see later), are automatically upgraded to the next kernel and photo stack about three to four months after the temporary release. 18.04.2 users will therefore see new packages from now on. All 19.04 applets and fixes back.

Most of the updates are:

- budgie-screen-applet
- budgie-weather-applet
- budgie-sys-monitor-applet
- budgie-cpufreq-applet
- budgie-advanced-brightness-controller
- Updated Pocillo theme

Ubuntu Budgie 19.04 will support for nine months until January 2020. If you need long-term support, it is recommended that you use Ubuntu 18.04 LTS instead.

Among the notes released on the covered areas are:

- New features and enhancements
- Problems solved
- Upgrades from 18.04/18.10 Ubuntu Budgie
- Known issues when developing
- Where to download Ubuntu Budgie

18.10 Features in 19.04

- Firefox is the default browser. So the chromium-browser icon pops out of both the icon – the task list and the dock – into Firefox.

- For a better rating, add a welcome budgie personality as a default program icon to make quick access (both to the thumbnail list and the wooden dock).

- We discarded TLP for the default installation. Saving power on the kernel from kernel 4.18 and later is important for new computers. TLP is still available for installation if kernel power saving does not affect you due to old CPU usage.

- Caffeine resource discarded – This replaces the caffeine budgie-applet that comes with the budgie-desktop v10.5.

19.04 Things to Look For

- New default background image and desktop.

- The fonts have changed from "Ubuntu" to "Noto Sans" without the terminal font, as Ubuntu is straightforward to read.

- Ubuntu Budgie Team Wallpapers 19.04 – The UB team has decided that this release will choose their favorite wallpapers – a lot of eclectic. I hope you like them. Next, we will run a 20.04 LTS community race.

- The translation team was busy – many more languages are now available.

- Browser Vote from the beginning now knows Snap – We offer GNOME Web and Midori as web browser options and Snap versions for Firefox and Chromium-Browser. Chrome, Vivaldi, and repo versions of Firefox and Chromium-Browser are still available.

- Catfish file and text search is now an automatic installation.

- The community has requested that the Files (Nautilus) be converted to Files (Nemo) – so you are also welcome to the partition screen and other great features. Tip – press Alt or right-click the toolbar to display the Favorite Menu, etc.

- Nemo can be launched on the Plank and the icon list applet.

- V3.8.6 is a Nemo version of the repository. V4.0.6 is available in our PPA backports.

- The nemo-drop box is available in our PPA backports.

- Nemo-share is available in our PPA backports – This adds a right-click option so folders can be shared without the need for root permission.

- Budgie-nemo integration – Right-click options to change the background, open budgie-desktop settings, and catfish file and text search.

- The color of the Nemo folder is unfortunately not found in the archives, so it is discarded.

- The latest version of moka-icon-theme is available – Pocillo uses Moka icons.

- Budgie-desktop is built on mutter 3.32 – so we benefit from the good running speed up the river.

- Introducing the default QogirBudgie theme from Vinceliuice Smart Theme – So three of the best default themes now in budgie-desktop settings (Arc, Pocillo, and QogirBudgie).

- The UB logo is proudly displayed on the About screen.

- The Pocillo-gtk theme has been upgraded – The visible key was the suggestion buttons. Green is now dark blue.

- Budgie-Welcome – We have removed our budgie-remix Reddit icons and the less-used G + page.

- Another Rhythmbox toolbar-integrated appmenu – Fix crashes with non-English locations.

Budgie 19.10 Released

The Budgie team has announced the release of 19.10 Eoan Ermine, the latest stable release of Ubuntu flavor featuring Budgie Desktop. Includes numerous developments from Ubuntu and Budgie Desktop projects. Ubuntu Budgie Eoan Ermine provides a better user experience with updated packages and tweaks.

The highlights of Ubuntu Budgie 19.10 are described below.

- Budgie Desktop integrated with the latest GNOME 3.34 stack. Also, the team has made significant contributions to making the overall desktop experience better and more consistent.

- Budgie Applets has been improved to work better with user experience. Advanced applets include window preview, instant char mini-app, vague clock applet, stopwatch applet, light control, status notification, etc.

- NVIDIA drivers are licensed in ISO. It allows the installation of NVIDIA drivers without an internet connection.

- Enables installation in ZFS partitions. It is a test feature, and Canonical tries to make this a significant factor in future LTS releases.

- Linux Kernel 5.3.

- Redesigned background selection from the GNOME 3.34 stack.

- LibreOffice 6.3.x.

Ubuntu Budgie 20.10

Version 21.10 of Ubuntu Budgie is supported for nine months until 2022 July. Long-term discharges such as 20.04.3 help stability (three years) compared to normal release for nine months.

Significant stability and refinement will be published in all supported sections of Ubuntu Budgie. We, the wider Ubuntu community, will initiate this and Canonical.

Some of the areas mentioned in this release note are as follows:

- Enhancements and new features

- Upgrades from version 20.04.3 or 21.04 version of Ubuntu Budgie

- Where to put Ubuntu Budgie

- Known issues when developing

- Small apps and apples

Mini Apps and Applets

- A few updated versions from many brilliant translators.

- Prepare to install spam on Syslog if the preview is enabled, but end users have entered the Wayland desktop session.

- Adjust the settings of the active Window Shuffler input system to work on nonlinear distribution.

- The Window Shuffler Control option has been completely redesigned and updated.

- Now, Window Shuffler contains an applet to reset and automatically move Windows, either a single window or a set of windows.

- Now, Shuffler also contains window rules.

- Shuffler Window Rules allow us to open application windows in some operating regions.

- The App Calculator Menu now displays a visual guide to copy the output to the site.

- Switching to Grid modes to the App Menu menu is now missing instead of slide animation.

- A few themes are placed on the category list in a strange way. We have more user-friendly coding within applet settings that allow us to control spaces.

- We no longer include a schema for basic gsettings. Ensures that we do not conflict with the installation of the pantheon. Now, the basic schema installation is found in the budgie apps schema.

- Changes to Budgie Calendar Applet: Properly manage the AM/PM appearance using custom date formats. It also returns to version 20.04 or 20.10.

- Changes to Budgie Calendar Applet: Select the current date when you open popover. It also returns to version 20.04 or 20.10.

- Changes to Budgie Calendar Applet: How and hide last week's option. It also reverts to versions 20.04 and 20.10.

- A version of Budgie-indicator-applet 0.7.0 has been published. It has configuration indexes that are used to display large icons.

- The new applet is the CPU temperature, which shows a few sensors' temperatures. This sensor was also shipped back in versions 20.04, 20.10, and 21.04.

BUDGIE'S DESKTOP

Version 10.5.3 of Budgie is a small edition explaining the many bug fixes, lifetime quality improvements, and GNOME 40 stack support.

1. Bug Repair and Cleaning

2. Budgie's Desktop

3. Raspberry Pi

4. Support for GNOME 40

5. Themes

6. Upgrades

7. Additional

Bug Repair and Cleaning

- Version 10.5.3 of Budgie describes several solutions for the Budgie panel applet, Window status tracking, and Raven. Drop Queue use was also canceled in Raven NotificationView. Instead of using any line, update the reference to NotificationWindow, as we are just worried about the header of the line to start with. We discarded Use Cancel usage of async pixbuf rating. There has not been any situation so far that we have not rated the time value icon before Notification Window is removed.

- Adjust the app icons before the Raven Notification Team title.

- Fix problems with the VirtualBox icon that can be displayed in the Applet, IconTaskList.

- Adjust the switch to the icon, that is, to manage Budgie Desktop Settings.

- Now, CanGoNext/CanGoPrevious, playback mode, and MPRIS metadata are done faster instead of being unnecessary.

- Now, notifications use the most important closing time.

- Now, the notification icon uses IconSize.DIALOG is more stable than IconSize.INVALID and only scales if the icon is not the required length.

Raven and Applet Audio Indicator

Use standard removal and mute, troubleshooting such as mute using media keys, and trying to mute with an applet that will not work before.

Raspberry Pi

Version 21.10:

- Compute Module Board Support 4.

- Add a command-line option, i.e., --force-arm-mode.

- Add a command-line option, i.e., --force-findpi-mode.

- Add a command-line option, i.e., --force-model modelname.

- Add a command-line option, i.e., --model modelname -cpuinfo "CPU String to Match."

- Improved access to a few options by using the mouse to move up and display the advice within the "status area."

- Disable buttons and simplify suggestions on the tab, i.e., Overclock if there is no "/boot/firmware/config.txt."

- Disable buttons and give advice on the tab, i.e., Show where pibootctl can be seen.

GNOME 40 Support

Version 10.5.3 version of Budgie Desktop specifies support for the GNOME 40 stack or in conjunction with GDM (GNOME login manager that benefits the gnome-shell).

In the settings of GNOME Daemon and GNOME Shell 40, there have been specific versions that affected Budgie's ability to use dedicated screens as a gnome-screensaver.

Upgrades

There are a few quality life upgrades to Budgie's version 10.5.3:

- The WM and Mutter schema settings are only used in the Budgie session.

- The full-screen status of the rewritten app for the use of window XIDs and diminished environments where a few unstructured flags will not be removed within the app will be removed as a full-screen application.

- Now, the space is adjustable within the status applet.

- There is an option in Budgie and revealed in the "Windows" section of Budgie Desktop Settings to automatically stop notifications when the window is in full-screen mode and does not rest when there is no full-screen window. It helps to minimize distractions when viewing or game content.

- The Lock Key applet may be pressed to toggle NumLock and CapsLock if xdotool is in the application.

Additional Information

- Now, Nemo Preview is available as a suggestion app in Budgie Welcome to visualize the content of any highlighted file. We need to click the space to convert the viewer into Nemo. It has also been restored to versions 20.04 and 20.10 again.

- Other Budgie Desktop Location fixes have been developed to make Budgie Desktop more friendly to other remote access points or desktop areas that use the dash as a default shell.

- In Budgie Welcome, IRC links have revealed that a Libera conversation from the Ubuntu family has ended its partnership with Freenode.

- The Drawing app has a downloaded version of 0.8.3.

- Now, 21.10 background images of UB teams are in everyday ISO.

UBUNTU BUDGIE 21.10 (IMPISH INDRI)

It is regularly published in the Ubuntu Budgie Group and is the official taste of the Ubuntu community containing the Budgie Desktop. This release will be valid for nine months until July 2022. Ubuntu Budgie uses Budgie Desktop v10.5.3 in this release and fixes the black theme with GTK + 4. In addition, windows Shuffler now automatically moves and configures windows and installs Windows 11 as the layout of your choice. Budgie 10.5.3 introduces stack support for GNOME 40.

Budgie 21.10 comes with Linux Kernel v5.13, Firefox v92 uses the native Debian package, while Ubuntu 21.10 uses the snap package, GNOME 40 applications, and other GNOME 41 applications, as well as -LibreOffice v7.2.1.

USING APT

Budgie Desktop is a modern and up-to-date Budgie Desktop that provides excellent desktop information and is available on all major Linux distributions. It is designed so that it can use a few system resources.

Budgie Desktop contains a specific application launcher called the Budgie menu that includes a few of our apps and is helpful. It provides an easy way to search and launch apps and features like the genre as you search and category-based filtering options. Also, Budgie contains a sidebar that can be accessed by pressing the icons on the far right inside the panel. We can see notifications, desktop control settings on the desktop, and a calendar.

We can get a few more features after installing and using Budgie Desktop. This section will show you how to download and install Budgie Desktop on Ubuntu.

Let's start with the installation process. There is no need to install a PPA in the Ubuntu operating system because Budgie is part of the official Ubuntu depot. Simply put, we will need to execute the installation process.

Step 1: We need to press the shortcut keys, that is, Ctrl + Alt + T, to start the terminal. After that, we will place an order below in our terminal to view updates:

```
$ sudo apt update
```

Step 2: Now, we will execute the command below on our Budgie Desktop Installation terminal:

```
$ sudo apt install personality-budgie-desktop
```

Step 3: After the installation process, notification will occur when selecting the display manager. We will choose lightdm and click the Install button.

It will take a few seconds to complete the process.

Step 4: We need to use the command below in our terminal to restart the system after the completion of the process:

```
$ sudo reboot
```

After the reboot, we will see the login screen below. We will press the icon under the desktop.

From there, select an option, that is, budgie-desktop.

We'll have a default look for Budgie's desktop.

We can now use Budgie's menu by pressing on the left side of the desktop.

If we no longer want Budgie Desktop and wish to return to our normal Ubuntu desktop, we will need to uninstall it by performing the following instructions on our terminal:

```
$ sudo apt to delete ubuntu-budgie-desktop ubuntu-
            budgie * lightdm
or $ sudo apt autoremove
```

To reinstall gdm3, use this command,

```
$ sudo apt install --reinstall gdm3
```

Installation of Budgie Using Tasksel

- We will use the tasksel command to install the Budgie Desktop. If a tasksel command is not available on your system, you can enter it by:

- $ sudo apt to install tasksel

- Use the following command to start Budgie Desktop installation:

- $ sudo tasksel install human-budgie-desktop

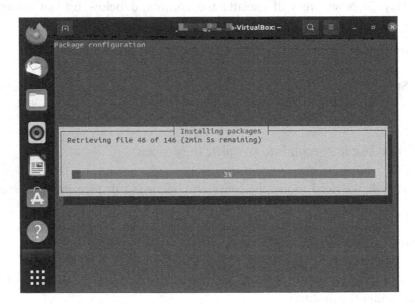

Budgie DE.

- Wait for tasksel to complete Budgie Desktop installation.

- Select lightdm configuration details to use TAB to select lightdm, and press the OK button.

- Restart your Ubuntu 20.04 system.

- Select a desktop session like Budgie and open the desktop selection menu.

- Select Budgie Desktop, enter your password, and press the login button.

NEW FEATURES AND DEVELOPMENT OF BUDGIE

Budgie Panel Dock Mode

Upstream Budgie ships have one panel automatically. It is uplifted and has an additional applet icon to manage your open applications.

While this is a handy set, it is not for everyone's preference or workflow. Ubuntu Budgie automatically sends the Plank dock app to its image. But while Plank is a good app, it won't be needed for long. The next update to Budgie adds a dock mode option that converts any panel on the desktop into a traditional style booth.

What Is a Plank Dock App?

One of our favorite Linux dock programs, Plank received a significant overhaul last month, adding much-needed features and healthy bug fixes.

Like other apps, Plank also has a theme, allowing you to create a tremendous seamless desktop experience.

Plank is not the only application like the dock available for Linux, but it is one of the best moving devices. It is well maintained, leaving the narrow view to system resources, and works well on all desktops and various window managers. It has a small but well-thought-out set of features and customization options. The heavier and more sophisticated instruments and whistles are not here. Instead, they are sold for reliability, stability, and simplicity.

Common Plank Features

- Displays "pinned" and active app icons

- Right-click the menu with quick menu options and items

- Drag dock items to rearrange

Options/settings include the following:

- Location: top, bottom, left, or right of the screen

- Align: collects icons to left, right, or center

- Behavior: which includes hiding options, e.g., "intelli-hide"

- Theme support: includes "transparent" option

Among the approximately 20 changes to Plank 0.8 are the following:

- The GTK theme "preferred background-color" was used for reference

- Favorites panel added

- Panel Mode object status adjusted

- The problem of the default folders in Linux Mint is fixed

- Delivery window delay increased to 200 ms

- Improved visibility on HiDPI screens

How to Install Plank Dock in Ubuntu

Plank 0.8 can install on Ubuntu 14.04 LTS or later (including outputs such as Linux Mint) via the official Docky PPA.

First, open a Terminal and paste the following instructions to add a PPA to your Software Sources, entering your password where instructed:

```
$ sudo add-apt-repository ppa: ricotz / docky
```

To update your repositories:

```
$ sudo apt-get update && sudo apt-get install plank
```

After installation, open the "Plank" app in Unity Dash (or the same App Menu in the desktop area you are using) and away! To unlock Plank preferences, you need to right-click on the "Plank" icon on the far left of the dock.

Intellihide

Budgie panels (including the new spiffy dock mode) take on additional coverage behavior. Among them is the famous intellihide feature, which causes the panel to hide cleverly when a window passes and removes a backup when it is released.

Apple of Night Light

It seems that every desktop that fits you adds a blue light filter to help prevent eye strain and promote better sleep. GNOME 3.24 has one (and more mobile), as do KDE Plasma, Windows 10 and (now) and ChromeOS. Eager to improve its users' eyesight and sleep habits, Budgie joins the fun. The next big release will see the desktop, including its new night light applet, full of on/off toggle, schedule options, and the ability to adjust color temperature.

New Budgie Settings App

As part of an effort to simplify Budgie's experience and allow for greater customization, Budgie moves the desktop settings out of the Raven sidebar. It incorporates them into its dedicated settings program. You will still be able to access standard features such as looks, fonts, and panel applets, but without showing them in a sidebar better suited to show you notifications and show useful widgets. If you have Eagle Eyes, you will notice many new features listed in the screenshot above, including the add-on section for managing auto startup applications and the customization option Budgie menu icon.

It Is Flexible

Using Ubuntu Budgie gives you the freedom to use the complete, fully integrated, pre-configured, if not all, of the applications you will need on your daily computer – or change anything in the way it looks, the way it works, or the apps you use to suit your preferences for you.

Built-in Security

Personality Budgie is built with safety in mind. Unlike apps that update only once a month, Ubuntu Budgie receives continuous updates. Updates include safety features for Ubuntu Budgie and all of its components. Security updates for all installed applications are also provided with the same schedule. This ensures that you have the latest protection for all your computer software, as soon as it is available! Powerful ApplicationsWhile Budgie Desktop provides essential visual user control and computer application. Ubuntu Budgie adds a collection of additional applications to transform your computer into a compelling workplace: from production to entertainment.

Games

A few years ago, Linux became the first-class playground, thanks to the arrival of Steam Platform on Linux. Steam has more than 2,000 high-quality indie degrees and AAA suitable for Ubuntu Budgie at the time of writing. While Steam is a big step forward to play on Linux, many high-quality and exciting open-source articles are available at Ubuntu Budgie. Whether you like airplanes, car racing, first-person shooters, jump and run, or card games, you will find something that will delight you.

Easy Migration

Connect your accounts, sync your calendars, and download your favorite apps. Ubuntu Budgie is your operating system.

CHAPTER 7

Cinnamon Desktop Environment

IN THIS CHAPTER

> ➤ Introduction

> ➤ History

> ➤ Software components

> ➤ User guide

> ➤ Cinnamon applications

> ➤ Installing Cinnamon on Ubuntu

> ➤ Core components and benefits

In the previous chapter, we discussed Budgie desktop. In this chapter, we will briefly discuss Cinnamon's desktop environment systems. Primarily, it is an official Linux OS and has various features. Cinnamon is the main desktop distribution platform for Linux Mint and is available as a desktop of choice for other Linux distributions.

DOI: 10.1201/9781003308676-7

INTRODUCTION

There are various terms to discuss to understand the concept of the Cinnamon. So let's begin with basic terms, then we move forward to the desktop environment Cinnamon.

Now we are going to cover basic terms before going deep into the Cinnamon desktop environment such as distribution, open-source Linux desktop environment, GUI, TUI, CLI, and so on.

What Is Distribution?

The term "distribution" refers to the combination of the packaging of the kernel with the GNU libraries and applications. Ubuntu is one such distribution. It contains the Linux kernel, the GNU tools, and many other applications and libraries.

Open-Source Linux Desktop Environment

The word "Open Source" is attributed to the Linux community which brought it into existence along with the introduction of Linux. "Linux" came into existence based on kernel. Many people and communities started contributing toward making it a complete operating system that could replace UNIX.

Key Points

- The freedom to run the program as per your wish

- Free software can be commercial

- The freedom to get the source code and make changes

- Legal considerations

- Contract-based licenses

Next, we will discuss GUI, CLI, and TUI, which are also related to the Ubuntu desktop environment Kool Desktop Environment (KDE).

This section examines the Graphical User Interface (GUI) and the significant components of the Linux GUI. You will learn about standard window managers and desktop environments used with Linux.

GRAPHICAL USER INTERFACE

GNU Network Object Model Environment (GNOME) is the default GUI for most Ubuntu installations and is (loosely) based on the Apple ecosystem. A GUI or graphical application is anything you can interact with using your mouse, touchpad, or touch screen. You have various icons and other visual prompts that you can activate with a mouse pointer to access the functionalities. DE provides the graphical user interface to interact with your system. You can use GUI applications such as GIMP, VLC, Firefox, LibreOffice, and file manager for various tasks.

Features of Linux GUI

The interface allows users to interact with the system visually with icons, windows, or graphics in a GUI. The kernel is the heart of Linux, whereas GUI is the face of the operating system provided by the X Window System or X.

The product of the X.Org Foundation, an open-source organization, X Window System is a protocol that allows links to be built on their X Server. You can use the X in one of the many window managers or desktop environments, such as the GNOME or the KDE. The desktop space includes a window manager and is a much more integrated system than a window manager. Built on a window manager, it requires both X Windows and a window manager.

Features of a GUI

There are unique features and tools to interact with the software to make the GUI as easy to use as possible. Below is a list of all of these with a brief description.

- **Button:** A graphical representation of a button that acts as a program when pressed.

- **Dialog Box:** The window type displays additional information and asks the user for input.

- **Thumbnail:** This is a small representation of a program image, feature, or file.

- **Menu:** A list of commands or options provided by the user through the menu bar.

- **Menu Bar:** It is a small, horizontal bar containing menu labels.

- **Ribbon:** Set up a file menu and toolbar that integrates program functions.

- **Tab:** A clickable area at the top of a window shows another page or location.

- **Toolbar:** The Button Bar, usually near the app window's top, controls software operations.

- **Window:** A rectangular section of a computer display that shows the operating system.

The GUI uses icons, windows, and menus to execute commands, such as opening, deleting, and moving files. Although the GUI operating system is navigated using the mouse, the keyboard can also use with keyboard shortcuts or arrow keys.

For example, if you wanted to open an application on the GUI system, you could move the mouse pointer to the system icon and double-click it. With the command-line interface, you will need to know the commands to go to the program's directory, enter the list of files, and then use the file.

Benefits of GUI

A GUI is considered more user-friendly than a text-based command-line interface, such as MS-DOS, or the shell of operating systems like UNIX.

Unlike command line or CUI operating systems, such as UNIX or MS-DOS, GUI operating systems are easy to read and use because commands do not need to be memorized. Additionally, users do not need to know any programming languages. Thanks to its ease of use and modern appearance, GUI operating systems dominate today's market.

Command-Line Interface

CLI is a command-line program that accepts inputs to perform a particular function. Any application you can use via commands in the terminal falls into this category. CLI is an old way of working with apps and applications and is used to perform specific tasks that users need. CLI is a text-based visual interface, unlike the GUI, which uses graphics options that allow the user to interact with the system and apps. CLI allows the user to perform tasks by entering commands. Its operating system is straight-forward but not easy to use. Users enter a command, press "Enter," and wait for a response. After receiving the command, CLI correctly evaluates it and displays the output/effect on the same screen. The command-line interpreter is used for this purpose.

CLI is introduced with a telephone typewriter. This system was based on batch processing. Modern computers support CLI, batch process-ing, and a single interface GUI. To make good use of CLI, the user must enter a set of commands (one by one) immediately. Many applications (mono-processing systems) still use CLI on their operators. In addition, programming languages like Forth, Python, and BASIC provide CLI. The command-line translator is used to use a text-based interface.

Another feature of CLI is the command line used as a sequence of char-acters used in the user interface or shell. Command information is used to inform users that CLI is ready to accept orders. MS-DOS is an example of CLI.

Terminal User Interface

TUI is also known as a Text-based User Interface. You have text on the screen because they are used only in the terminal. These applications are not well-known to many users, but there are a bunch of them. Terminal-based web browsers are an excellent example of TUI programs. Terminal-based games also fall into this category. Text User interface (also known as written user interaction or terminal user interaction) is a text-based user. TUIs differ from command-line communication in that, like GUIs, they use all of the screen space and do not provide line-by-line output. However, TUIs use only the text and symbols found in the standard text terminal, while GUIs typically use high-definition image terminals.

CINNAMON

It is a free and open-source desktop X Window System source from GNOME 3, following standard desktop metaphor agreements. Cinnamon is the main desktop distribution platform for Linux Mint and is available as a desktop of your choice for other Linux distributions and other appli-cations such as Unix.

The development of Cinnamon began as a reaction to the April 2011 release of GNOME 3 when the standard desktop GNOME 2 desktop was left in favor of GNOME Shell. Following several attempts to extend

GNOME 3 to suit the design goals of Linux Mint, Mint developers have installed several GNOME 3 components to create a standalone desktop space. The split on GNOME was completed on Cinnamon 2.0, released in October 2013. Apples and desktops are no longer compatible with GNOME 3.

As a distinguishing feature of Linux Mint, Cinnamon has generally received good media coverage, mainly due to its ease of use and soft learning curve. In terms of its sequential design model, Cinnamon is similar to the Xfce desktop and GNOME 2 desktop (MATE and GNOME Flashback).

History

Like a few other GNOME-based desktop environments, including Canonical Unity, Cinnamon became dissatisfied with the GNOME team's traditional desktop experience in April 2011. Until then, GNOME had included a standard desktop theme. But GNOME 3 was replaced by GNOME Shell, which lacked a function-like panel and other basic desktop features. The abolition of these essential features was unacceptable for distribution developers such as Mint and Ubuntu, aimed at users looking for collaborative sites where they could be relieved immediately.

The Linux Mint team initially decided to upgrade the GNOME Shell extensions to overcome this difference to replace the remaining features. The results of this effort were "Mint GNOME Shell Extensions" (MGSE). At the time, the MATE desktop environment was also forked from GNOME 2. Linux Mint 12, released in November 2011, then merged both, thus giving users the choice of GNOME 3-with-MGSE or traditional GNOME desktop 2.

However, even with MGSE, GNOME 3 was still missing out on the luxury of GNOME 2 and was not well received by the user community. Some missing features could not be changed at the time, and it seemed that the extensions would not work in time. In addition, GNOME developers could not meet the needs of Mint developers. To give Mint developers better control over the development process, GNOME Shell was forked as "Project Cinnamon" in January 2012.

Gradually, various key applications were replaced by Mint developers. As of version 1.2, released in January 2012, the Cinnamon window manager is Muffin, the fork of GNOME 3's Mutter. Similarly, from September 2012 (version 1.6 onwards), Cinnamon includes a Nemo file manager with a fork from Nautilus. Cinnamon-Control-Center, installed since May 2013 (version 1.8 onwards), integrates GNOME-Control-Center functionality with Cinnamon-Settings and makes it possible to manage update applets, extensions desktops, and themes. -top. GNOME-Screensaver was also installed with a fork and is now called Cinnamon-Screensaver.

As of October 2013 (version 2.0 onwards), Cinnamon is no longer a front-end desktop for a GNOME desktop like Unity or GNOME Shell but a desktop environment that is unique in its own right. Although Cinnamon is built on GNOME technology and uses GTK, it no longer requires GNOME to be installed.

Overview

The Cinnamon desktop environment is a vast development project. Between 2006 and 2010, the most prominent Linux Mint desktop space was GNOME 2. It was very stable and trendy. In 2011, Linux Mint 12 could not be shipped via GNOME 2. The upcoming GNOME team released a brand-new desktop (GNOME 3 aka "GNOME Shell") using the latest technology (Clutter, GTK3), which had a completely different design, and we used a very different paradigm than the previous one but used the same word spaces and therefore could not be installed near GNOME 2. After the decision from Debian to upgrade GNOME to version 3, GNOME 2 was no longer available in Linux Mint. To address this issue, two new projects were launched.

A project called "MATE" was started by an engineer named Perberos. Its goal was to redesign and repack GNOME 2 to its original form. Linux Mint started a project called "MGSE." Its goal was to upgrade the GNOME 3 extensions to replace some of the lost and found functionality in GNOME 2 (panel, systray, application menu, alt-centric tab selector, window-list, etc.).

Linux Mint 12 has been distributed on both MATE and GNOME3 + MGSE. Six months later, and after a huge amount of work, MATE was stable, and from the extension set, MGSE became a GNOME 3 fork called Cinnamon. Linux Mint 13 was the first Linux release for the Cinnamon desktop deployment. Since Linux Mint has a MATE and Cinnamon system, both provide users with a desktop, one forked from GNOME 2 and the other forked and based on GNOME 3.

Other recent improvements include the following:

- Desktop grid, wildcard support for file search, multi-process settings daemon, desktop actions on panel launcher, various desktop management processes, and file manager in Nemo.

- An additional option for the desktop panel layout offers a modern-looking theme with customized windows.

- Improved design of duplicate applications in the menu (i.e., flatpak vs. deb packages) pin files in Nemo focuses on performance improvement.

SOFTWARE COMPONENTS

X-Apps

Cinnamon introduces X-apps based on GNOME core applications but is modified to work across Cinnamon, partner and xfce; they have a standard user interface (UI).

- Xed is a Gedit/pluma text editor.

- Xreader is a document viewer based on Evince/Atril.

- Xviewer is an image viewer based on Eye of GNOME.

- Pix is an image editor based on gThumb.

- Xplayer is a media player based on GNOME (Totem) videos.

Features

Features provided by Cinnamon include:

1. Desktop effects, including animation, transition effects, and transparency using make-up

2. Panels with the main menu, launchers, window list, and system tray can be adjusted left, right, the top, or bottom edge of the screen

3. Various extensions

4. Apples from the panel

5. An overview of activities similar to those at GNOME Shell

6. Settings editor for easy customization:

   - Panel

   - Calendar

   - Themes

- Desktop results

- Apples

- Extensions

The volume and light adjustment are applied to the scroll wheel while pointing to the taskbar icon. As of January 24, 2012, there were no official documents for Cinnamon itself, although most GNOME Shell documents apply to Cinnamon.

USER GUIDE

Amazing GUI

It is no secret that GUI seems essential to any Linux distribution. We even have users who like a particular distro just because of the beautiful nature of the desktop, and Cinnamon is one of them. The Cinnamon desktop does not come with unnecessary icons and shortcuts.

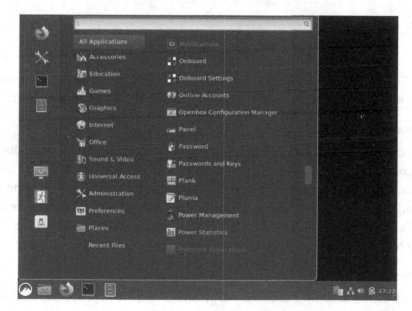

Cinnamon Desktop.

You can stop using System Settings and go to the desktop menu. You can choose whether they appear in the primary or secondary monitor or all user monitors with multiple monitors.

Use of Desklets

Cinnamon offers several desklets, including a weather app, a desktop photo frame application, time and date, sticky notes, CPU, or disk monitors.

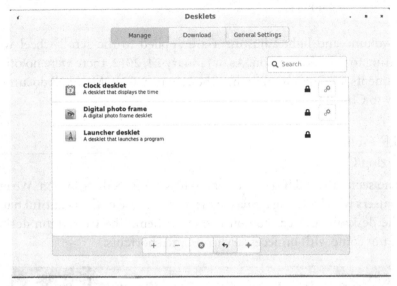

Adding Desklets.

One of my favorites is the Time and Day desktop. That's what we call desklets – a one-purpose app that adds to your desktop for easy access.

Speed

Since you started using Linux, some have used Cinnamon on Fedora, Debian, and Ubuntu. Cinnamon worked well in all of these distros and quickly and easily launched apps. The desktop itself does not take long to load once it is entered. Also, Cinnamon is compatible with hardware on PCs with fewer resources like Ram. Unless you use additional applications such as Desklets, Cinnamon should work fine on your PC.

Integration

With Linux desktop locations, you will find specific applications in them all. For example, you say you want to install a GNOME-screenshot in another location like KDE, Mate, or Budgie – if the specified location does not support the required GNOME screen libraries, you will not be able to install it. Fortunately, with Cinnamon, things are getting a lot better because of the strong integration of libraries. You can install multiple applications regardless of the desktop environment for which they are designed. Cinnamon supports all KDE libraries, GNOME, and other Desktop domains.

Extremely Customizable

Although not available as KDE, Cinnamon allows you to customize it enough to have a good user interface. The control center application gives you access to all Cinnamon desktop configurations. You can open a window in the configuration feature in the window that opens. You can also change the look of your desktop image using theme options in system settings. You can customize window parameters, thumbnails, and pointers, and upgrade your desktop. Additionally, you also customize the fonts and desktop background.

All visibility settings are located at the top of the window. Everything in the "System Settings" window looks clean and tidy. Go to System Settings and find something similar to that given below.

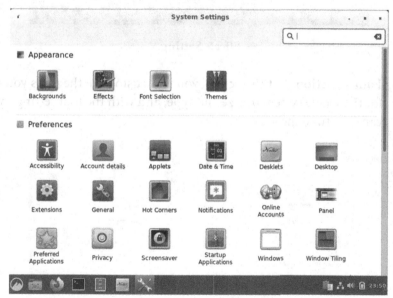

System Settings.

There are various options available in the column, such as effects, appearance, font selection, and themes. Let's take a look at these.

- **Effects:** The effects options are simple, self-explanatory, and straightforward. You can turn the effects of various desktop items on and off or change the window conversion by reversing the effects style. If you want to change the speed of results, you can do it with the customization tab.

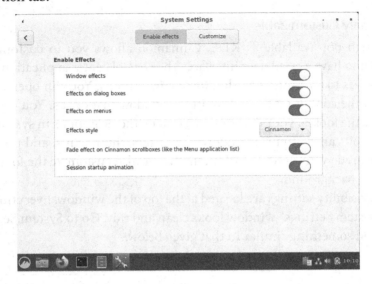

Effects Setting.

- **Font selection:** In this section, you can customize the fonts you use for the whole system by size and type, and with the font settings, you can fine-tune the look.

Font Selection.

- **Themes and symbols:** Users of Linux Mint do not need to go everywhere to change what they want. Window manager, icon, and customization panel are all in one place. You can change your panel to black or light color and window settings to suit your changes. The default Cinnamon settings look great on eyes, and they used the same when checking out the Ubuntu Cinnamon Remix but in orange color.

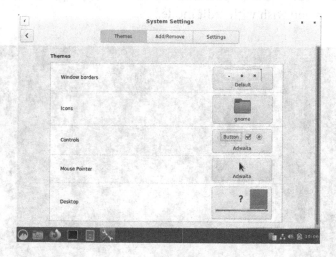

Themes and Icons.

- **Customize what is on your desktop screen:** The background is not the only desktop item you can change. You can find more options by right-clicking on the desktop and clicking "Customize."

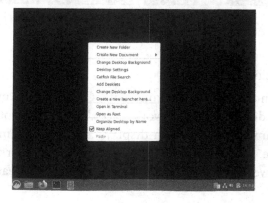

Customize Settings.

Cinnamon Panel

The Cinnamon panel or toolbar comes with a very simple setup: the open application menu, system tray, and application selector. However, you can customize the panel and add new program launchers. To do this, go to the program menu, right-click on the program you want, and select "Add to panel." Additionally, you can add launchers to the desktop or the "Favorites" bar. Cinnamon also allows you to set program launchers on the panel as you wish with "edit mode."

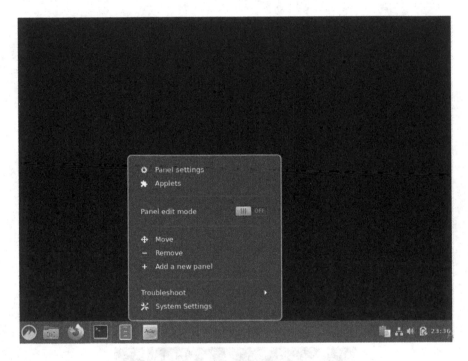

Adding Panel.

Use of Multiple Workplaces

Like other Linux desktop environments, Cinnamon allows you to use only multiple desktops called "workspaces." You select your workspaces on the Cinnamon panel, which displays a window frame throughout the workspace. A nice feature with Cinnamon in the workspace is moving apps between workspaces or assigning them to each workspace available.

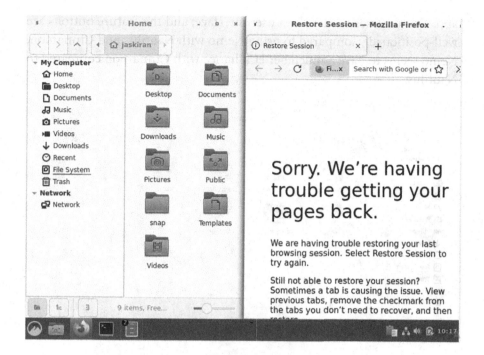

Multiple Workspaces.

Flexibility

Cinnamon desktop environment makes it easy to use and navigate due to the high flexibility in the Graphical User Interface. For example, if you use only a few applications, it can be challenging to get reduced applications or those running in the background from the toolbar. One of the features you can use to solve this is to reset the operating system buttons to a selector. This way, you can easily click on any operating system without much hassle. Also, Cinnamon desktop has a helpful menu that gives you access to a few features and utilities, including desklets, create a new document, and more.

Nemo

Like any other desktop environment, Cinnamon has its fair share of automated apps. Another one that caught my eye was the file manager, Nemo. You will have worked closely with other file managers like Nautilus and Krusader. With Cinnamon, that's when you get exposure to this great app. Nemo comes with a clean user interface, and the feature buttons are well-positioned. Compared to using Nemo with Nautilus and Cinnamon, Nemo seems to have a much better alliance with Cinnamon compared to Nautilus.

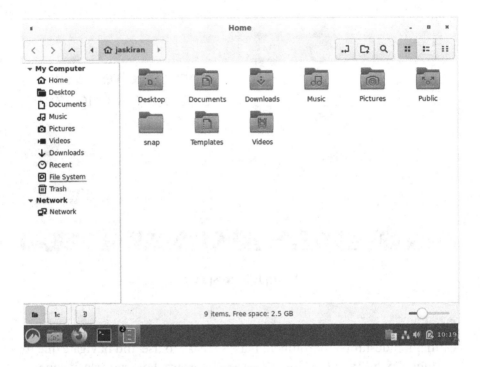

Nemo File Manager.

Stability

As of October 2013, Cinnamon has ceased to be a GNOME frontend over GNOME and no longer needs GNOME to be installed. Share desktop space with its rights. Since then, the developers have added many features, and as of this writing, the latest release is Cinnamon 3.6.7. Cinnamon is very stable and works well with several Linux distributions.

Cinnamon Apples

Cinnamon apples are all the features in your panel below, such as a calendar or keyboard layout switch. In the Administration tab, you can add/remove pre-installed applets. You should check out the apples you can download. The weather and CPU Indicator applets were my choices in addition.

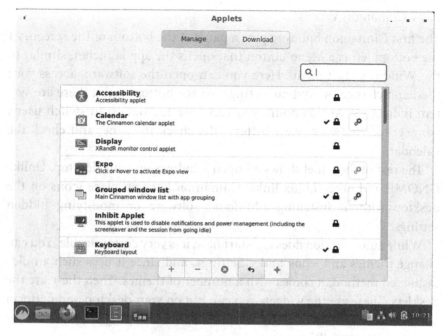

Applets.

Desktop Wallpaper

To change the desktop background on the Cinnamon desktop, right-click on the desktop and select "Change Desktop Background." It will open an easy-to-use window, where the system folders are available on the left side, and in the mounting window, there is a preview of photos within each folder.

CINNAMON APPLICATIONS

Cinnamon introduces X-Apps based on GNOME Core Applications but modified to work across Cinnamon, MATE, and XFCE; they have a standard user interface.

- Xed (text editor)

- XviewerAUR (image viewer)

- (Document viewer) xreader

- XplayerAUR media player

- PixAUR image editor

How Cinnamon Works

The first Cinnamon building puts a panel at the bottom of the screen. At the bottom left is a Menu button that opens the app launcher, similar to the Windows Start menu. Here you can open the software, access your files, and change the system settings. At the bottom right, there are system indicators. At this point, you can have the option to switch users, connect to a network, view battery life, check the time, and check the calendar.

The rest of the panel shows all open windows on your desktop. Unlike GNOME and other Linux links, Cinnamon lets you place icons on the desktop without installing additional software or modifying hidden settings.

While auto-creation does not start new, it is very customizable. You can change themes and icons under Settings, and since it uses such an old-fashioned method, it comes with a number of themes. Then there are the desklets. These are the widgets you can put on your desktop and perform simple tasks, such as showing you the weather, keeping a quick note, or monitoring your CPU usage.

INSTALLING CINNAMON ON UBUNTU

There are some steps that are slightly different for Ubuntu 20.04 and the older LTS version 18.04.

```
$ sudo apt install cinnamon
```

Or

```
$ sudo apt-get install cinnamon
```

When you run the above command in the terminal, you will see the following lines of code:

```
username@username-virtualbox:~$ sudo apt-get install
cinnamon
[sudo ] password for username:
Reading package lists... Done
Building dependency tree
Reading state information... Done
The following additional packages will be installed:
```

```
cinnamon-common cinnamon-control-center
        cinnamon-control-center-date
cinnamon-common-goa cinnamon-control-screensaver
        cinnamon-control-center
```

The Cinnamon package is available at the entire Ubuntu 18.04 repository. However, it only uses the version of Cinnamon 3.6 as you see using the apt show command. There used to be an official PPA from the Cinnamon group of Ubuntu, but it no longer exists. Do not give up. There is an illegal PPA available, and it works perfectly.

This PPA contains a version of Cinnamon 4.2, which is not the latest but better than 3.6. Open the terminal and apply the following instructions:

$ sudo add-apt-repository ppa: embrosyn / cinnamon

$ sudo apt update && sudo apt to add cinnamon

It will download files up to 150 MB in size. This also provides you with Nemo (Nautilus fork) and Cinnamon Control Center. These bonus features give a Linux Mint experience.

After installing, you will see your installed desktop environment as given below:

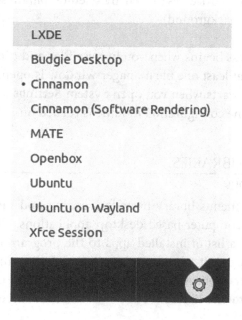

Cinnamon Desktop Environment.

Removing Cinnamon from Ubuntu

Understandably, you may want to take out Cinnamon. First, switch back to GNOME any desktop space you used before.

Now, delete the Cinnamon package:

```
$ sudo apt to remove cinnamon
```

If you have used PPA to install Cinnamon, you should also remove the PPA from your list:

```
$ Sudo add-apt-repository -r ppa: embrosyn / cinnamon
```

After logging in, the following processes are started automatically:

- Cinnamon-session (session manager that starts all other processes)

- Cinnamon (the visible part of the Cinnamon desktop)

- Nemo-desktop (with desktop icons and desktop content menu)

- Cinnamon-screensaver (screensaver)

- various csd- * processes (which are daemon plugin settings and running in the background)

The Nemo process begins when you browse files and directories. It stays open as long as at least one file manager window is open. The Cinnamon settings process starts when you open System Settings and stays open as long as at least one configuration module is turned on.

CINNAMON LIBRARIES

Cinnamon-Menus

The Cinnamon menus library provides reading and monitoring activities for a set of computer-based desktop applications. Cinnamon menus can quickly add a list of installed apps to the program menu, download menu system icons, alt-tab selector, and window list, and keep this data synced whenever applications are installed or removed from a computer. The Cinnamon menus library is built in C, and the source code is available on Github.

Here are some resources currently available on Cinnamon-desktop:

Resources	Description
nnamon.desktop	dconf settings schemas used by several Cinnamon components
libcvc	A PulseAudio utility library used to control sound volume and devices
Gnome-rr	An Xrandr utility library to detect, load, and save monitor configurations
gnome-xkb	A keyboard layout utility library
gnome-bg	A wallpaper utility library
gnome-installer	A cross-distribution library used to install software applications

This document explains how to create a user-friendly application category, usually displayed as a menu. It allows third-party software to add menu items to all desktops and allows system administrators to edit menus in a way that affects all desktops.

Cinnamon-Desktop

It is a collection of libraries and settings used by other parts of Cinnamon. Whenever multiple desktop components need access to the same app (whether a setting or a proper function), we put this device on a Cinnamon desktop.

Cinnamon-desktop contains libCinnamon desktop library, Cinnamon-about, the program and other comprehensive desktop documents. The lib-Cinnamon-desktop library provides a shared API for several applications on the desktop, but that can't sit on a multi-platform platform for reasons. There is no API or ABI guarantee, though we do our own best to provide stability. API documents are available via gtk-doc.

Muffin

Muffin, or libmuffin, is more accurate in the window handling library. The Windows Manager does not work with a separate process within the Cinnamon desktop area. The primary Cinnamon process uses the libmuffin library and uses both visual components (panel, applet, etc.) and a window manager.

Clutter

Clutter and Cogl libraries are part of the muffin package now. It is a library for creating and displaying both 2d and 3d image features. It is used by both the muffin itself (e.g., mixing and setting the stage) and St in Cinnamon (all St widgets are clutter players). Cogl is a library used for clutter in 3d rendering. Muffin is built in C, and source code is available on Github.

CJS

CJS is a Cinnamon Javascript translator. It uses Mozilla's SpiderMonkey, which makes it possible to interact with GObject and connect with the GIR, GNOME, and Cinnamon libraries using that language. CJS is also processed within a significant Cinnamon process, and desktop components written in Javascript are contained in a large portion of Cinnamon. CJS has been upgraded to C ++ and Javascript, and the source code is available on Github.

CORE COMPONENTS

Cinnamon-Session

The Cinnamon-session manager is responsible for presenting all the required components after logging in and closing the session properly if you want to exit. Among other things, the session manager introduces the key components required for the session (such as the desktop itself and its components) and applications configured to start automatically.

The Cinnamon-session also provides a DBus interface called the Existing Interface Connector, making it easy for applications such as media players to set busy times and prevent power management (pause, long sleep, etc.) and storage-screen during video playback. Session managers allow applications to register to be closed automatically. For example, a text editor is registered for a session when it is launched and works with it is on the go. If the document is not saved, the session recognizes it and allows you to save your work before signing out.

Cinnamon-Settings-Daemon

The Cinnamon-settings-daemon is a collection of processes that run in the background during your Cinnamon-session. The Cinnamon-settings-daemon provides a wide range of sessions and functions requiring a long process. Among the services used by the Cinnamon-settings-daemon are XSettings Manager, which provides themes, font, and other settings to GTK + programs, and the clipboard manager, which stores the clipboard's contents when the program exits. Many user interface features of

Cinnamon and Cinnamon settings depend on the Cinnamon-settings-daemon in their functionality.

The Cinnamon-settings-daemon takes the name org.Cinnamon.SettingsDaemon into the session bus to ensure that only one event is active. Some plugins submit items under this name to make their functionality available to other applications. Communication of these items should be considered confidential and unstable.

Here is a description of some of these processes.

Processes	Description
csd-automount	It automatically mounts hardware devices when they are plugged in
csd-clipboard	It manages the additional copy-paste buffer available via Ctrl+C/Ctrl+V
csd-housekeeping	It handles the thumbnail cache and keeps an eye on the space available on the disk
csd-keyboard	It handles keyboard layouts and configuration
csd-media-keys	It handles media keys
csd-xsettings	It handles X11 and GTK configuration
csd-xrandr	It handles screen resolution and monitors configuration
csd-wacom	It handles wacom devices
csd-print-notifications	It handles printer notifications
csd-power	It handles battery and power management
csd-orientation	It handles accelerometers and screen orientation
csd-mouse	It handles mice and touch devices
csd-mouse	It handles mice and touch devices
csd-mouse	It handles mice and touch devices

Cinnamon-Screensaver

Cinnamon screensaver is responsible for the lock screen and the minimal handling of certain power management functions (although most of them are csd-enabled within the Cinnamon Settings Daemon). Cinnamon-screensaver is built into Python, and source code is available on Github.

Cinnamon

The Cinnamon github project is the largest and most active project within the entire project.

It contains various sub-sections labeled C:

- **St:** It is a Cinnamon's widget toolkit written on top of Clutter.

- **Appsys:** It is an abstraction of Gio.AppInfo and Cinnamon-menus, providing metadata on installed applications.

- **DocInfo:** It is an abstraction of recently opened documents.

- **Tray:** It is a small library for managing status icons.

The visible layer of the desktop is written in Javascript:

- **Cinnamon JS:** The panels, window management, HUD, effects and most of what you see

- **Applets:** An applets is within the panel

- **Desklets:** The desklets is on top of the desktop

System settings, configuration modules, and scripts are written via Python. Upgraded to C, Python, and Javascript and source code is available on Github.

Nemo

Nemo is Cinnamon's file manager. When you open your home directory or browse files using Nemo, another small part of Nemo is the Nemo-desktop. Its role is to manage desktop icons and desktop content menu. When you log in, the Nemo-desktop starts automatically with a Cinnamon-session. The Nemo process only starts when you browse the directions and stops when you open the last file manager window. Nemo is upgraded to C, and source code is available on Github.

Nemo Extensions

Nemo offers a set of APIs and is very easy to expand, both in C and Python. Nemo extensions is a Github project where common extensions are stored. Some Nemo extensions are made in C and some in Python. Their source code is available on Github.

Cinnamon Control Center

Although Cinnamon settings (part of the Cinnamon project) and most of its modules are written in Python, a few activation modules are still being written in C. Today, only a few modules are located in the Cinnamon-control-center:

- Color profiles

- Date and time configuration

- Show display configuration and monitoring

- Network configuration

- Online accounts configuration for online accounts

- Wcom device configuration devices

- Cinnamon-control-center is upgraded with C, and source code is available on Github

X-APPS

A project called "X-Apps" was launched in 2016 to produce standard GTK desktop applications. The idea for this project is to replace applications that are no longer integrated well outside of the area (this is the case with the growing number of GNOME applications) and provide our desktop locations with the same set of key applications, so that each change, each new feature upgrade, and each minor improvement will not gain only one place, but all.

The main ideas of X-Apps are as follows:

- Using modern tools and technologies (GTK3 HiDPI support, settings, etc.)

- Using standard user interface (titlebar, menubar)

- To work everywhere (be it normal, desktop-agnostic, or distro-agnostic)

- To provide performance to users who already enjoy (or have enjoyed in the past by distributing lost functionality)

- To keep up with the reverse (to work on as many distributions as possible)

All three Linux Mint programs come with the same XApps libraries and applications. When working on XApps, our development efforts are focused on improving all desktops.

- **libxapp:** This is the XApps library. Anything cross-desktop goes in there. Available in Python and JS as well, via GObject Introspection. This project was developed in C, and its source code is available on Github.

- **python-xapp:** This is a small Python library that offers additional functionality. This project has been developed in Python, and its source code is available on Github.

- **xed:** Xed is based on Pluma and works as a default text editor.

- **xviewer:** It is based on Eye of GNOME and works as a default image viewer.

- **xplayer::** It is based on Totem and works as a standard music and video media player.

- **xreader:** It is based on April and works as a default document and PDF reader.

- **pix:** It is based on gThumb, an app that organizes your photos.

- **blueberry:** Bluetooth device, blueberry, is ahead of GNOME-bluetooth with systray support. The GNOME Bluetooth frontend has been removed from GNOME-bluetooth and has become part of the GNOME-control center, making GNOME-bluetooth useless without GNOME. Blueberry provides a non-existent frontend and makes it easy for some GTK desktops to use GNOME-bluetooth.

- **greetings:** Slick-greeter is an automatic login screen, a LightDM priest that was originally forked from uninity-greeter and configured to run itself (except GNOME-settings-daemon, GNOME or unity).

- **lightdm-settings:** The lightdm settings project provides a setting tool to set up LightDM and slick-greeter.

BENEFITS

It is still distributed under the terms of the GNU General Public License. The basic technology is made with a fork from the GNOME desktop. Since version 2.0, Cinnamon is a complete desktop and not just a GNOME location like GNOME Shell and Unity. Cinnamon is a fast version of 3D, which should be used frequently. It provides users with an "easy-to-use and comfortable desktop experience" while staying up-to-date with the latest technology and power. The DE was founded by Clement Lefebvre, who also developed and continues to develop the Linux Mint distribution for Linux. Linux Mint has also collaborated on the development of a GNOME desktop-based desktop platform known as MATE. It is the main Desktop for Linux Mint distribution and is available as a desktop of your choice for other Linux distributions and other applications such as UNIX.

Extension

Cinnamon can be replaced with themes, applets, and extensions. Themes can customize the look of Cinnamon features, including but not limited to the menu, panel, calendar, and box usage. Apples are icons or text from a panel. Five applets are shipped automatically, and developers are free to create their own. A tutorial on creating simple apples is available. Extensions can change the performance of Cinnamon. Developers can upload their themes, applets, and extensions to the Cinnamon webpage and allow users to download and rate.

Flexibility

Cinnamon Desktop has an excellent exit menu that you can access with the right-click. This menu contains a selection of other commonly used functions such as accessing Desktop Settings and adding Desktop, as well as other Desktop-related functions.

Consolidation

The choice of Desktop does not depend on the availability of long-term applications. All apps, regardless of the Desktop they are designed for, will work well on any other desktop, and Cinnamon is the same. All libraries required for the use of applications written for KDE, GNOME, or any other desktop are available and make using any Cinnamon desktop application a seamless experience.

Speed

Cinnamon is quick and easy. Applications upload and display instantly. The Desktop itself loads quickly during login, although this is my personal experience and is not based on any timely testing.

Visible Connector

Cinnamon has a clean look using easy-to-read fonts and color combinations. The Desktop is not interrupted by unnecessary rooting, and you can edit what icons are displayed on the Desktop using System Settings => Desktop Menu. This menu also allows you to specify whether desktop icons are displayed only on main monitors, only on secondary monitors, or on all monitors.

Discovery

Cinnamon is available in Linux Mint 12 storage and is included in all versions of Linux Mint 13 and above as one of the four possible desktop environment options, the other being MATE. It is also the user of your choice in the Linux Mint Debian Edition Update Pack 4 respin.

Settings Editor

Settings editor for easy customization. Customize panel, calendar, themes, desktop effects, applets, and extensions.

Overview Mode

New views all over the world have been added to Cinnamon. These two methods are "Expo" and "Scale," which can be configured in Cinnamon Settings.

Audio Enhancements

Audio settings are completely rewritten in Python to provide additional features such as notifying users when no input device is available on the custom page, visual changes in the audio settings panel, the ability to mute all volume controls simultaneously, with additional adjustments to the volume slider. / mute buttons, check the audio dialog boxes, and add the appropriate icons.

Touchpad Settings

Cinnamon users can gain full control of the touchpad pad available on their previously unlocked portable computer. The center-click action is fixed in the image frame.

CHAPTER SUMMARY

In this chapter, we have covered the introduction of Cinnamon with its features, history, core projects, applications, and development. Also, we have a separate section where you get a history of Cinnamon versions. Cinnamon has features that DEs like GNOME and Unity do not have. Cinnamon is custom-made for DE and does not require any external plugin, widget, or tweak tool to customize the desktop. Cinnamon can benefit any new Linux user with fantastic and necessary features.

CHAPTER 8

LXDE Desktop Environment

IN THIS CHAPTER

- ➤ Introduction history
- ➤ Core components
- ➤ Installation of LXDE in Ubuntu
- ➤ Reasons to use LXDE
- ➤ Lubuntu distribution

In the previous chapter, we discussed Cinnamon. In this chapter, we will briefly discuss the Linux-based operating system LXDE. Firstly we will learn the fundamentals of LXDE.

INTRODUCTION

LXDE (abbreviation for Lightweight X11 Desktop Environment) is a lightweight desktop operating system like Unix and Unix, such as BSD and Linux. It is designed to use fewer system resources than other desktop environments like GNOME. Initially approved by Taiwanese program editor Hong Jen Lee in 2006, it was written in C programming language using the GTK tool kit; the new version, released in July 2013, is based on Qt.

DOI: 10.1201/9781003308676-8

The popularity of LXDE has grown slightly since its first release and is now the default desktop distribution platform for several Linux, including Lubuntu, Knoppix, and Raspbian. Its tests have shown that it requires about half as much RAM compared to operating systems. It also uses less power, making it a good choice for portable computer users who want to extend their battery life.

Overview

LXDE is written in C programming language, using the GTK 2 tool kit, and works on Unix and other POSIX compliant platforms, such as Linux and BSD. The LXDE project aims to provide faster and more powerful desktop space. In 2010, experiments suggested that LXDE 0.5 had the lowest memory usage of the four most popular desktop devices (GNOME 2.29, KDE Plasma Desktop 4.4, and Xfce 4.6). It consumed less power suggesting that Linux-distributed mobile computers use LXDE 0.5 and discharge their batteries slower than those at other desktop locations.

LXDE uses output extracts from its components (or groups that have a corresponding dependency). The window manager (default) used is Openbox but can configure a third-party window manager with LXDE, such as Fluxbox, IceWM, or Xfwm. LXDE combines the GPL licensed code with the LGPL licensed code.

After installing the basic Gentoo system and the X server, there are several options to consider in terms of which site has the best images to use. Many options are available, from minimalistic window managers such as Openbox to desktop environments like KDE and GNOME. Some users may want a lightweight image area but do not wish to install and configure each component, like Openbox. For some time, users in this position have been installing Xfce because it fits the definition of a lightweight area and comes with a suite of light applications. While Xfce provides a fully integrated environment without the extensive use of the KDE or GNOME application, it can rely on the heavy side. Finally, another Xfce version has been created: Lightweight X11 Desktop Environment or LXDE for short. Users, however, who do not like all those dependencies (dbus, polkit, etc.) which are pulled by lxsession should check out Lumina.

HISTORY

The LXDE was started in 2006 by Taiwanese program coordinator Hong Jen Yee, also known as PCMan, when he published PCManFM, a new file manager and the first LXDE module. Compared to DistroWatch's

Linux distribution rates in early January 2011, Ladislav Bodnar noted an increase in LXDE's popularity compared to other desktop environments. He said that

> The exciting thing is the increase in distribution that uses a light-weight LXDE desktop but is full of features or an Openbox window manager if you look at the tables. For example, Lubuntu is beating Kubuntu well in page hits, while CrunchBang Linux, a lightweight distribution via Openbox, is still in the top 25 despite failing to produce stable releases for more than a year. Many other distributions began offering LXDE-based programs for their products, which contributed to the dramatic increase in the popularity of this new desktop space.

Qt Port

LXDE is heading to Qt in 2013 and is trying to work with Razor-qt, another lightweight desktop. Hidden characters in the application menu are in traditional Chinese. On July 3, 2013, Hong announced the full Qt port of the LXDE suite, and on July 21, Razor-qt and LXDE announced plans to integrate two projects. This integration meant that GTK and Qt versions would be available for a long time, but, in the end, all the original team efforts were focused on the Qt port, LXQt.

CORE COMPONENTS

There are various core components of Qt port in LXDE as given below.

Window Manager (Openbox)

The desktop environment was developed by "PCMan" in Taiwan, aimed at soft and beautiful code and usable design. It uses Openbox as a Window Manager, which has parts of the lightweight module.

Openbox is a free X Window System window manager, licensed under the GNU General Public License (GPL). Openbox was initially based on Blackbox 0.65.0 but has been completely rewritten in the C programming language, and version 3.0 is not supported on any code from Blackbox. It is designed to be minor, fast, and fully compliant with the Inter-Client Communication Conventions Manual (ICCCM) and the Extended Window Manager (EWMH) Tips. It supports many features such as menus where the user can control applications or display various dynamic information. The author of Openbox is Dana Jansens of Carleton University in Ottawa, Ontario, Canada.

Usage

Openbox allows right-clicking (or another component) "root menu" on desktop and enables users to configure windows management. When the window is lowered, it becomes invisible. To highlight windows, use the Alt + Tab or desktop menu, which is accessible by right-clicking (or, again, any other user-binding object) menu. Extending Openbox and other small programs that add icons, task layers, launchers, eye candy, and more like driver restore are common.

Features

The Openbox menu system has a way of using flexible menus. It is done by accepting the output of the text and using the output as a menu source. Each time the user points his mouse to the submenus, the script is restarted, and the menu is restarted. This capability allows users and software developers to have more flexibility than the usual menus available from other window managers.

File Manager (PCManFM-Qt)

PCMan File Manager is a manager application developed by Hong Jen Yee of Taiwan, intended to replace GNOME, Dolphin, and Thunar files. PCManFM is a standard file manager in LXDE, also developed by the same author, in collaboration with other engineers. Since 2010, PCManFM has completely rewritten it from the beginning; building, setup, and configuration instructions have changed in the process.

It is issued under the GNU License; PCManFM is free software. It follows the information provided by Freedesktop.org for collaboration. Dissatisfied with GTK3, Hong Jen Yee tested Qt in early 2013 and released the first version of PCManFM based on Qt on March 26, 2013. He clarified, however, that it does not show any form of GTK to LXDE, saying: "GTK and Qt versions will be there together." The new PCManFM-Qt is an integral part of the LXQt. By 2020, the Arch Linux community installed PCManFM on GTK 3.

Features

- Full support for GVfs with seamless access to remote file systems
- Twin panel

- Thumbnail images

- Desktop Management – displays a background image and desktop

- Bookmarks

- Many languages

- Browsing tabs (such as Firefox)

- Volume management such as mount/unload/unload, requires gvf

- Drag and drop support

- Files can be dragged between tabs

- File organization (Default application)

- Provides the following views: thumbnail, merge, detailed list, thumbnail, and left-sided scrub

Desktop Panel (razor-panel)

Razor-qt is a free and open desktop environment for open source. It was intended as a lightweight desktop space based on the Qt application framework, and was "designed for users who appreciate simplicity, speed, and a clear visual interface." The development of Razor-qt is interrupted, as it is integrated with the LXDE Qt port to make LXQt.

Razor-qt was still in its early stages of development. As of February 2012, the environment included panel and switch viewer, desktop, app launcher, settings center, and sessions. These components may be enabled or disabled by the user.

Razor-qt can be used with any modern X window manager such as Openbox, fvwm2, or KWin. Razor-qt memory usage was slightly higher than LXDE, using 114 MiB in the reviewer test, while LXDE used 108 MiB.

Merge with LXDE

After LXDE developer Hong Jen Yee sent PCManFM to Qt in early 2013. Yee and other interested developers negotiated a possible partnership with Razor-qt, another open desktop platform with similar software design principles. The first release of the new product, LXQt v0.7.0, was made public on May 7, 2014. The final release of razor-qt was on January 12, 2013.

Session Manager (LXSession)

LXSession is a standard session manager used by LXDE. LXSession automatically starts a set of applications and sets up an active desktop. Additionally, it does not work on a desktop and can be used by any window manager.

Merge Flags

- "--open-man": Produces personal pages.

- "--enable-more-warnings": Enables alert for additional integration during construction.

- "--enable-gtk3": Includes GTK3 when partially compatible (incomplete).

- "--enable-buildin-clipboard": Add built-in clipboard support, using GTK2.

- "--enable-buildin-polkit": Add building support within the polkit agent (based on GTK).

- "--enable-debug": Allows for debugging.

- "--enable-gtk": Allows GTK + applications and integration. It can pass --disable-gtk build without any component of GTK +.

Applications and Binary

1. Lxclipboard: An application to allow clipboard support using GTK

2. lxlock: Screen lock application, using external applications

3. Lxpolkit: Polkit agent

4. Lxsession-default: Wrap in Dbus mode to launch applications described in the lxsession configuration file

5. Lxsession-default-apps: A lxsession configuration application (especially for debugging purposes)

6. lxsession-edit: An old lxsession configuration application

7. Lxsession-utils: Mixed resources for lxsession

8. lxsettings-daemon: Xsettings daemon

Policykit Agent (razorqt-policykit)

Razor-qt is an advanced, easy-to-use, and fast desktop environment based on Qt technology. It is designed for users who appreciate simplicity, speed, and an accurate visual interface. Unlike most desktop environments, the Razor-qt also works well with less powerful machines.

Display Manager (sddm or Lightdm+razorqt-lightdm-greeter)

SDDM is the default display manager that appears when you first log into Lubuntu or exit or select another session. Simple Desktop Display Manager (SDDM) is the display manager (image entry program and session manager) for X11 and Wayland window systems. SDDM was written from scratch in C ++ 11 and supported themes with QML. SDDM is a free and open-source software subject to GNU General Public License version 2 or later. LXQt developers recommend SDDM as display manager.

Features

- Supports different display technologies (X11 and Wayland using Mir)

- Supports remote login (incoming – XDMCP, VNC, outgoing – XDMCP, connected)

- Comprehensive test suite

- Compliance standards (PAM, login, etc.)

- It has well-defined interface between the server and the user interface

- Cross-desktop (greeters can be written in any tool kit)

- A well-defined greeting API that allows multiple GUIs

- Support for all conditions of use of manager-display, and plug-in where appropriate

- LightDM has a simpler code base than GDM and does not load any GNOME libraries to operate, but at the expense of other features the user may or may not need them.

COMPONENTS OF LXDE SOFTWARE

Unlike other large desktop areas like GNOME, LXDE components have fewer dependencies and are not tightly integrated. Instead, they can be installed independently or in LXDE itself.

Components	Descriptions	Notes
LXInput	Mouse and keyboard configuration tool	It is a small program used to configure keyboard and mouse for LXDE. It requires intltool 0.40.0 and GTK+ 2.12 or newer
PCMan File Manager	File manager and Desktop metaphor provider	PCMan File Manager is a manager application developed by Hong Jen Yee of Taiwan, intended to replace GNOME, Dolphin, and Thunar files. PCManFM is a standard file manager in LXDE, also developed by the same author, in collaboration with other engineers. Since 2010, PCManFM has completely rewritten it from the beginning; building, setup, and configuration instructions have changed in the process
LXLauncher	Easy-mode application launcher	LXLauncher is an optional launcher for program enhancements on 7 "-10.2" screens LXLauncher enables the LXDE desktop to be a theme-oriented theme for collecting specific app applications in a single view under tabs
LXPanel	Desktop panel	LXPanel is a standard LXDE panel. The desktop panel can generate a menu of automatically installed applications from *. desktop files. It can be configured from the GUI preferences dialog, so editing configuration files is unnecessary. The section provides a "Run" box with auto-complete
LXSession	X session manager	LXSession is a standard session manager used by LXDE. LXSession automatically starts a set of applications and sets up an active desktop. Additionally, it does not work on a desktop and can be used by any window manager
LXAppearance	GTK theme switcher	LXAppearance is the standard theme switcher of LXDE. Users are able to change the theme, icons, and fonts used by applications easily. Starting at version 0.6. 1, it also allows to enable the accessibility features
GPicView	Image viewer	lxde.org. GPicView is a standard LXDE image viewer. GPicView incorporates instant lightning launching and precise interface

(Continued)

Components	Descriptions	Notes
LXMusic	A frontend for the XMMS2 audio player	LXMusic is a small LXDE music player. It is a visual function (GUI) of xmms2 audio player, lightweight and integrates with server/client design
LXTask	Task manager	LXTask is a standard task manager and LXDE system monitoring. It starts with Ctrl + Alt + Del and is very simple. To delete an unresponsive app, you can use Term or Kill
LXRandR.	A GUI to RandR	LXRandR is a standard LXDE screen manager. Controls screen adjustment and external monitors. This is a very basic setup tool that uses the X RandR extension. Front GUI command line program xrandr. LXRandR can let you change the screen setting on the go
LXDM	X display manager	LXDM is a lightweight display manager that aims to replace gdm in LXDE distros. UI is used with GTK +. In the meantime, LXDM uses a gtk pixmap engine to complete its theme, but in the future, it will use a gtk-css-engine that will free up theme development
LXNM	Lightweight network connection helper daemon Supports wireless connections (Linux only)	LXNM is a standard LXDE network communication server daemon. Sets the wireless connection as fast as possible. Currently, it only works with Linux. They are no longer under development, and most LXDE packages or all distros use NetworkManager
Leafpad	Text editor	Leafpad is an open source text editor for Linux, BSD, and Maemo. Created with the focus of being a lightweight text editor with minimal dependencies, it is designed to be simple and easy-to-compile. Leafpad is the default text editor for LXDE Desktop environment, including Lubuntu up to version 18.04 LTS. After Lubuntu moved to the LXQt desktop, Leafpad was replaced by FeatherPad
FeatherPad	Text editor	It is a free text editor available under GPL-3.0-or, the latest license. Developed by Pedram Pourang of Iran, written in Qt, and works on FreeBSD, Linux, Haiku OS, and macOS, Lubuntu used the Leafpad text editor as part of the GTK-based LXDE desktop
ObConf	A GUI tool to configure Openbox	Obconf, Openbox GUI setup editor. There are only two configuration files found in ~ /. config / openbox. They are called menu. xml and rc

(Continued)

Components	Descriptions	Notes
Xarchiver	File archiver	Xarchiver is the default archive for lightweight desktop environment like Xfce and LXDE. The features and capabilities of this archive manager are not the same as the archive manager, but you are able to manage popular archive file formats such as ZIP, TAR and RAR

INSTALLATION OF LXDE IN UBUNTU

LXDE is a lightweight desktop alternative of Unity, GNOME, and KDE. It is suitable for older computers and anyone who wants a faster, lighter system. It is much simpler than Xubuntu's XFCE. LXDE contains the basic features of open but accessible desktop space. It doesn't have a lot of shiny graphics or unnecessary features that block you.

Launch the following command to install the custom LXDE for Ubuntu and vanilla LXDE on Ubuntu system.

```
$ sudo apt-get install lubuntu-desktop
```

Here is the example of installation of Lubuntu given below.

```
username@username-virtualbox:~$ sudo apt-get install
                lubuntu-desktop
[sudo ] password for username
Reading package lists... Done
Building dependency tree
Reading state information... Done
The following packages additional packages will be
                installed:
2048-qt qpt-config-icons-large
                apt-config-icons-large-hidpi
Apt-xapian-indec ark bluedevil breeze-cursor-theme
                catdoc cdparanoia
```

Use this command instead of adding vanilla LXDE only:

```
$ sudo apt-get install lxde
```

You can also have an option by downloading Lubuntu CD live. Lubuntu comes from Ubuntu with LXDE installed automatically.

The first screen of the LXDE is:

LXDE Desktop.

Current Developments

Despite the team moving to LXQt development, some developers have continued to keep LXDE on GitHub, and, as of March 2021, there is a new commitment to supporting the latest version of GTK 2. From July 2019, the LXTerminal release is based on GTK 3 to avoid relying on the old VTE lib.

GTK 3 Port

As of May 2020, there is a GTK 3 test port built by the Arch Linux community. GTK 3 versions are already designed for the following components: LXAppearance, LXAppearance-ObConf, LXDE-common, LXDE-icon-theme, LXDM, LXhotkey, LXInput, LXLauncher, LXPanel, LXRandR, LXSession, LermSession, PCerminal, and Open. Another advantage of using GTK 3 is that the GTK 3 systems operate traditionally in Wayland. PCManFM is a popular file manager that can be used with tile window managers, so having PCManFM native Wayland is helpful for people who use Sway.

Default Desktop

- Knoppix

- LXLE Linux

- Peppermint OS

- Raspberry Pi OS

- Trisquel Mini

- Arch Linux

- Artix Linux

- Debian

- Devuan

- Fedora

Let's discuss each of the following in brief.

KNOPPIX

It is a Debian-based operating system designed to run directly on a CD/DVD (Live CD) or USB flash drive (Live USB), the first of its kind in any application. Knoppix was founded, and named after him, Linux mentor Klaus Knopper. It is loaded from the removable drive and then pressed into a RAM drive when you start the program. Decompression is noticeable and on the way.

Although KNOPPIX is primarily used as a Live CD, it can also be installed on a hard disk as a standard application. Computers that support the launch on USB devices can download KNOPPIX from a live USB flash drive or memory card. There are two main programs: standard compact-disc (700 megabytes) and DVD (4.7 gigabytes) "Maxi" edition. The CD edition has not been updated since June 2013 until recently. From version 9.1, CD images are also released. Each major program has two language programs: English and German.

KNOPPIX usually contains free and open-source software, but it also includes some proprietary software, as long as it meets certain conditions. Knoppix can easily copy files to hard drives with inaccessible applications. To quickly and safely use Linux software, Live CD can be used instead of installing another OS.

More than 1,000 software packages are included in the CD edition, and more than 2,600 packages are included in the DVD program. Up to 9 GB can be stored on DVD in compressed format. These packages include:

- LXDE, a lightweight X11 desktop environment; default from Knoppix 6.0 and later

- MPlayer, with MP3 audio and Ogg Vorbis audio playback support

- Internet access software, including KPPP dialing and ISDN services

- Iceweasel Web Browser (based on Mozilla Firefox)

- Icedove Email Client (based on Mozilla Thunderbird)

- GIMP, an image fraud program

- Data recovery and system optimization tools

- Network analysis and management tools

- LibreOffice, complete office

- Terminal server

Minimum requirements for Knoppix hardware:

- Intel/AMD-compatible processor (486 or later)

- Minimum RAM requirements:

- 32 MB of text mode

- Live location without modification:

- 512 MB in image mode with just LXDE

- 1 GB to use web browser and production software

- Recommended 2 GB

- Bootable Optical Drive:

- DVD-ROM for current versions

- CD-ROM version 7.2 and higher, or launch floppy and standard CD-ROM (IDE/ATAPI or SCSI)

- Standard SVGA compatible graphics card

- Serial or PS/2 is a standard mouse or USB mouse compatible with IMPS / 2

LXLE Linux

LXLE is a Linux distribution based on Ubuntu/Ubuntu LTS, using the LXDE desktop environment. The LXLE is a lightweight distro focused on visual aesthetics, working well for old and new hardware.

Peppermint OS

It is a Debinu Stable-based Linux distribution using the Xfce desktop. The Peppermint OS is a Debinu Stable-based Linux distribution, using the Xfce desktop environment. It aims to provide a standard setting for newcomers to Linux, which requires low hardware resources to work.

Peppermint OS was initially conceived at the Black Rose Pub in Hendersonville, NC (North Carolina), the USA, at night drinking and discussions about the future of the Linux desktop. Peppermint was initially designed to be a social media platform.

The construction of the alpha development included possible guidelines before the decision to install Lubuntu was made. There was a little experimentation with KDE, E17, Adobe Air, and several code bases between January and February 2010. Alpha was built using the Lubuntu 10.04 code base starting in March 2010. Peppermint was released for a small group of independent beta testers.

Releases

Peppermint One was released On May 9, 2010, and it received more than 25,000 downloads. It quickly passed its webmaster and switched to VPS. NET. VPS.NET became the first sponsor of the Peppermint project. Peppermint Ice was released on June 20, 2010. Play Chromium as the default browser and install a blue and black theme to separate it from Peppermint One.

Peppermint Two was released on June 10, 2011, features from two previous programs, Packing Chromium as its default browser next to the Ice program for creating Direct Site Browsers. It was also the first Peppermint version in both 32 and 64-bit versions.

On July 23, 2012, Peppermint Three was released. Chromium stable storage is automatically enabled; specific theme and fixed artwork; a few default web applications in the menu; posted via GWoffice; and GIMP 2.8 added to the Peppermint repository.

On June 13, 2013, Peppermint Four was released. Peppermint Four was based on code Ubuntu 13.04 and used the LXDE desktop space with Xfwm4 instead of Openbox as a window manager. Model games, Entanglement, and First Person Tetris have been added. Added are meta-packages of famous works such as photography and photography in the Featured Software Manager section.

On June 23, 2014, Peppermint Five was released.

> With this release, we are preparing for the future. The state of technology is constantly changing, and we are constantly responding to meet the needs of our users. We are 100% driven to deliver a fast, secure, and widely available OS. Peppermint Five is another step forward.
>
> Shane Remington, COO of Peppermint OS, LLC

On May 31, 2015, Peppermint Six was released.

> Peppermint is pleased to announce the launch of our latest operating system, Peppermint Six. Simple and built-in speed, Peppermint Six offers promise whether you use the software on your desktop, online, or using cloud-based applications. I want to take this. It is an opportunity to thank Mark Greaves, who has stood and produced much of what you see here at Peppermint Six. Mark now plays a major role here at Peppermint by leading the development team. I think he will be impressed with what he and others put together at Peppermint Six.
>
> Shane Remington, COO of Peppermint OS, LLC

On June 24, 2016, Peppermint Seven was released.

> The Peppermint team is pleased to announce our latest Peppermint 7 application, coming out of both 32bit and 64bit systems and the newest version with the full support of UEFI / GPT / Secure Boot, a new version of Ice (our framework Specific Browser Home Host) is also fully supported by Firefox and Chromium / Chrome web browser.
>
> Mark Greaves, Peppermint, 2016

On January 14, 2020, Peppermint chief executive Mark Greaves (PCNetSpec) died. After taking Peppermint to Shane Remington and Kendall Weaver, shortly after Peppermint Five, Mark dedicated his life to Peppermint with the support of his family and went on to release more versions, up to Peppermint 10 and respin for Peppermint 10. Official announcement was made. At Peppermint forum, a memorial fund was set up by his family to honor Mark's legacy.

On February 2, 2022, PeppermintOS released a new version for the first time in two years, with its major new features and changes including:

- Peppermint is now based on Debian Stable 64-bit, instead of Ubuntu or its derivative

- Reduced LXDE components instead of Xfce

- Nemo replaces Thunar as a default file manager

- No web browser installed. The browser can be installed using the Welcome to Peppermint application

- Ubiquity has been replaced by Calamares system installer

Raspberry Pi OS

The Raspberry Pi OS (formerly known as the Raspbian) is a Debian-based Raspberry Pi program. Since 2015, the Raspberry Pi Foundation has officially provided as the primary operating system for the Raspberry Pi family of single-board computers. Mike Thompson and Peter Green created the first Raspbian version as a standalone project. The first building to be completed was completed on July 15, 2012.

Raspberry Pi OS is a highly developed Raspberry Pi line for single-board integrated computers with ARM CPUs. It works on all Raspberry Pi except Pico microcontroller. Raspberry Pi OS uses the modified LXDE as its desktop component with an Openbox packaging window manager and unique themes. Distribution is accompanied by a copy of the algebra program Wolfram Mathematica, Minecraft version called Minecraft: Pi Edition (note that Minecraft: Pi Edition is no longer included as a Debian bullseye update), and a lightweight version of Chromium web browser.

Versions of Raspberry Pi

Raspberry Pi OS has the following installation options with two 64-bit options:

- Raspberry Pi OS Lite 32-bit and 64-bit

- Raspberry Pi OS 32-bit and 64-bit

- Raspberry Pi OS Full 32-bit

- Raspberry Pi OS Lite is a tiny version and does not include a desktop environment

- Raspberry Pi OS installs Pixel Desktop Environment

- Raspberry Pi OS Full comes pre-installed with additional production software

All versions are still distributed as .img disk image files. These files can flash to microSD cards where Raspberry Pi OS is running. In March 2020, the Raspberry Pi Foundation published the Raspberry Pi Imager. This custom disk allows the installation of Raspberry Pi OS and other Raspberry Pi-enabled applications, including RetroPie, Kodi OS, and others.

Features

1. **User Interface:** The Raspberry Pi OS desktop, PIXEL, looks similar to most common desktops, such as macOS Microsoft Windows, and is based on LXDE. The menu bar is set up and contains an application menu with Terminal, Chromium, and File Manager shortcuts. A Bluetooth menu, a Wi-Fi menu, a volume controller, and a digital clock are to the right.

2. **Package Management:** Packages can be installed using APT, the Recommended Software application, and the APT GUI cover through the Add/Remove Software tool.

3. **Components:** PCManFM is a file browser that allows quick access to all computer locations and is redesigned for Raspberry Pi OS Buster (2019-06-20). Raspberry Pi OS initially used Epiphany as a web browser but switched to Chromium by launching its redesigned desktop. Raspberry Pi OS comes with many basic IDEs, such as Thonny Python IDE, Mu Editor, and Greenfoot. It is also shipped with educational software such as Scratch and Bookshelf.

4. **Trisquel:** Trisquel is a computer application, a Linux distribution, available in another distribution, Ubuntu. This project aims for a

free software program without patent software or firmware and uses a version of Ubuntu's modified kernel, with non-free code (binary blobs) removed. Trisquel relies on user contributions. Its logo is the triskelion, a Celtic symbol. The Free Software Foundation lists Trisquel as a distribution containing only free software.

History

The project started in 2004 with Vigo University's support for Galician language for educational software. It was officially launched in April 2005 with Richard Stallman, founder of the GNU Project, as a special guest. According to project director Rubén Rodríguez, Galician's support has sparked interest in South American and Mexican communities of immigrants from Ourense Province. In December 2008, Trisquel was added by the Free Software Foundation (FSF) to its Linux-approved list of Free Software Foundation distributions.

Versions

Five basic versions are available, as given below.

1. Trisquel

2. Trisquel Mini

3. Triskel

4. Trisquel Sugar TOAST

5. Trisquel NetInstall

Let's have a look at each of the following.

- **Trisquel:** The standard Trisquel distribution includes MATE desktop environment and user interface (GUI), English, Spanish, and 48 other local, 50 in total, in 2.6 GB live DVD image. Some translations can be downloaded if an internet connection is available at installation.

- **Trisquel Mini:** The Trisquel Mini is one of the main Trisquel, designed to work well on older netbooks and hardware. It uses LXDE and low GTK + and X Window System resources which are different from other GNOME and Qt-KDE applications. The LXDE desktop-only includes local English and Spanish descriptions and can consist of a 1.2 GB live DVD image.

- **Triskel:** Triskel is another major Trisquel using the KDE interface, available as a live 2.0 GB ISO DVD.

- **Trisquel Sugar TOAST:** Sugar is a free and open desktop space designed for use by children in shared learning. Sugar replaces the standard MATE desktop available with Trisquel.

- **Trisquel NetInstall:** NetInstall contains a 25MB CD image with just a small amount of software to execute a text-based network installer and download the remaining packages online.

There are various other alternate desktops for LXDE, as given below.

1. **Arch Linux:** It is a Linux distribution with x86-64 processors. Arch Linux adheres to the KISS policy ("Keep It Simple, Stupid"). The project strives to have minimal distribution-related changes. As a result, it reverses a bit with updates, makes more sense than design ideas, and focuses on customization rather than ease of use. Pacman, a package manager, written for Arch Linux, is used to install, uninstall, and update software packages. Arch Linux uses an outgoing model, which means there is no "major release" of entirely new versions of the system; a regular system update is all that is required to get the latest Arch software; installation photos released monthly by the Arch team are the latest images of critical system components.

2. **Artix Linux:** Artix is an Arch Linux-based output distribution that uses OpenRC, runit, s6, suite66, or dinit init instead of systemd. Artix Linux has its package repositories but, as a Pacman-based distribution, it can use packages from Arch Linux archives and any other exit distribution, even packages depending on systemd. Arch User Repository (AUR) can also be used. The Arch OpenRC started in 2012, and the Manjar OpenRC was developed alongside it. In 2017, these projects were put together to build Artix Linux.

3. **Debian:** Debian has access to online storage containing more than 51,000 packages. Debian officially includes the only free software, but non-free software can be downloaded and installed on Debian repositories. Debian has popular free programs such as LibreOffice,

Firefox web browser, Evolution mail, K3b disc burner, VLC media player, GIMP image editor, and Evince document viewer. Debian is a popular choice for servers, such as part of the LAMP stack operating system.

Features

1. A few flavors of Linux kernel are available in each port. For example, the i386 port tastes IA-32 PCs that support Local Extension and real-time computers, older PCs, and x86-64 PCs. Linux kernel does not contain firmware without resources, although such firmware is available for free packages and other installation media.

2. Debian offers CD and DVD images specially designed for XFCE, GNOME, KDE, MATE, Cinnamon, LXDE, and LXQT. MATE is officially supported, while Cinnamon support is added with Debian 8.0, Jessie. Unusual window managers like Light, Openbox, Fluxbox, IceWM, WindowMaker, and others are available.

3. Multimedia support has been a problem for Debian regarding codecs threatened with possible copyright infringement, without resources, or under limited licenses. Although packages with distribution problems may enter the uncomfortable area, libdvdcss is not hosted on Debian.

REASONS TO USE LXDE

1. LXDE supports multiple panels. Like KDE and Cinnamon, LXDE game panels contain a system menu, app launchers, and an activity bar displaying active app buttons. The first time you logged into LXDE, the panel setup looked incredibly familiar. LXDE seems to have taken over the KDE configuration of my favorite top and bottom panels, including system tray settings. The app launchers on the top panel seem to come out of the Cinnamon configuration. Panel content makes it easy to present and manage programs. By default, there is only one panel under the desktop.

2. Openbox Setup Manager provides a single, simple tool for managing the look and feel. Provides theme options, window decorations, multi-monitored window behavior, window movement and resize, mouse control, desktops, and more. It is much more complicated

than setting up the KDE desktop. Openbox offers a fantastic amount of control.

3. LXDE has a robust menu tool. Tere is an exciting option that you can access in the Advanced tab of the Desktop Favorites menu. The long name for this option is, "It shows the menu provided to window managers when the desktop is clicked." Once this dialog box is selected, the Openbox desktop menu is displayed instead of the standard LXDE desktop menu when right-clicking desktop. The Openbox desktop menu contains almost every optional menu you want, and everything is easily accessible on the desktop. Includes all app menus, system management, and favorites. It also has a menu containing a list of all the simulated apps installed so sysadmins can easily present their favorite.

4. By design, its desktop is clean and straightforward. Nothing can stop you from getting your work done. Although you can add clutter to the desktop in the form of files, directory folders, and application links, no widgets can be added to the desktop. You may like some widgets on my KDE and Cinnamon desktops, but it's easy to integrate and needs to move or minimize windows or use the "Show desktop" button to clear the entire desktop.

5. LXDE comes with a solid file manager. The default LXDE file manager is PCManFM, which became my file manager during my time with LXDE. It is flexible and can work better for most people and situations. PCManFM allows multiple tabs to be opened by right-clicking on any item in the sidebar or by left-clicking on the new tab icon. The Locations window to the left of the PCManFM window displays the applications menu, and you can launch applications on PCManFM. The upper part of the Locations window also displays the device icon, which can be used to view your storage items, a list of removable devices and buttons that allow you to mount or lower them, as well as the Home, Desktop, and Trash folders for easy access. The bottom section of the Places panel contains shortcuts for other default guides, Docs, Music, Photos, Videos, and Downloads. You can also drag additional references to the Shortcuts section of the Places window. The Zones window can be replaced with a standard guide tree.

6. The title bar of the new window lights up when opened behind existing windows. This is a great way to get new windows for many existing ones.

7. Xfce Power Manager is a small, powerful application that lets you configure how power management works. It provides General Configuration tab and System, Display, and Devices tabs. The Devices tab displays a table of devices attached to my systems, such as mighty battery mice, keyboards, and UPS. Displays information about one, including the vendor and the serial number, and the battery charge status.

LUBUNTU DISTRIBUTION

It is a Ubuntu-based lightweight Linux distributor and uses the LXQt desktop instead of GNOME desktop. Lubuntu was initially managed as "light, resource-efficient, and energy-efficient," but now aims at "efficient but standard distribution focused on getting out of the way and allowing users to use their computer." Lubuntu initially used the LXDE desktop, but moved to the LXQt desktop with Lubuntu 18.10 in October 2018, due to slow LXDE development, loss of GTK 2 support, and being active, and stable LXQt development without GNOME reliance. The name Lubuntu is a portmanteau for LXQt and Ubuntu. The name LXQt is derived from the combination of the LXDE and Razor-qt project, while the word Ubuntu means "humanity for others" in Zulu and Xhosa. Lubuntu received official recognition as an official member of the Ubuntu family on 11 May 2011, starting with Lubuntu 11.10, released on October 13, 2011.

History

The LXDE desktop was first made available to Ubuntu in October 2008, with the release of Ubuntu 8.10 Intrepid Ibex. These earlier versions of Lubuntu, including 8.10, 9.04, and 9.10, were not available as separate downloads for ISO images and can only be installed on Ubuntu as different lubuntu-desktop packages from Ubuntu repositories. LXDE can also be added to previous versions of Ubuntu.

In February 2009, the LXDE project became a self-care project within the Ubuntu community leading to a new official release called Lubuntu. In March 2009, the Lubuntu project was launched on Launchpad by Mario Behling, which included the logo for the first project. The project established an official Ubuntu wiki project page, listing applications, packages, and components.

In August 2009, the first ISO test was released as a Live CD, with no installation option. A preliminary test in September 2009 by Linux Magazine reviewer Christopher Smart showed that Lubuntu's RAM was almost part of Xubuntu and Ubuntu in standard installations and desktop applications, with at least two-thirds of live CD usage.

In 2014, the project announced that GTK +-based LXDE and Qt-based Razor-qt would be integrated with the new Qt-based LXQt desktop. The transformation was completed with Lubuntu 18.10 in October 2018, the first regular release of the LXQt desktop lease. Lenny became Lubuntu's masquerade in 2014.

During the 2018 transition to LXQt, Lubuntu's goal was re-imagined by the development team. It was initially intended for users with older computers, usually ten years or older. Still, with the introduction of Windows Vista PCs, older computers gained faster processors and more RAM, and by 2018, ten-year-old computers were mainly left. As a result, the Lubuntu development team decided to shift the focus to emphasizing well-written, LXQt, "providing users with practical but modular information," which is not automatically automated and available in any language. Developers also decided to stop recommending minimum system requirements after the release of 18.04 LTS. In August 2018, Lubuntu 20.10 would automatically switch to the Wayland display server protocol. In January 2019, the developers formed the Lubuntu Council, a new organization to establish their former organization, with its written constitution.

LIST OF APPLICATIONS

Lubuntu LXDE versions include the following applications:

User Apps

- **Abiword:** word processor
- **Sounds:** a music player
- **Evince:** PDF reader
- **File-roller:** archiver
- **Firefox:** a web browser
- **Calculator:** calculator
- **GDebi:** package installer
- **GNOME Software:** Package Manager
- **Gnumeric:** spreadsheet

- **Guvcview:** webcam
- **LightDM:** login manager
- **Light Locker:** screen lock
- **MPlayer:** video player
- **mtPaint:** graphics
- **Pidgin:** instant messenger and microblogging
- **scrot:** screenshot tool
- **Simple scanner:** scanning
- **Sylpheed:** email client
- **Synaptic Software Center:** package managers
- **Transfer:** bittorrent client
- Update Manager
- **Startup Disk Creator:** USB ISO author
- **Wget:** command line web downloader
- **XChat:** IRC
- **Xfburn:** CD burner
- **Xpad:** recognition

From LXDE

- **GPicView:** image viewer
- **Leafpad:** text editor
- LXLooks
- LXDE Common
- LXDM
- LXLauncher
- LXPanel
- LXRandr

- LXSession

- Edit LXSession

- LXShortCut

- LXTask

- LXTerminal

- Menu-Cache

- **Openbox:** window manager

- **PCManFM:** file manager

Up to and including the 18.04 LTS, Lubuntu was also able to access Ubuntu storage through Lubuntu Software Center, Synaptic package manager, and APT that allows installation of any applications available on Ubuntu.

CHAPTER SUMMARY

In this chapter, we have covered LXDE with its features, along with history, core projects, applications, version history. We have a separate section where you get a history of LXDE versions.

Other Desktop Environments

In the previous chapter, we discussed the desktop environment LXDE. The chapter discussed information such as history, installation, features. In this chapter, we will learn about three desktop environments – LXQt, Enlightenment, Pantheon – and their history, versions, and some useful features.

LXQT DESKTOP ENVIRONMENT

LXQt is a free and open-source lightweight desktop. It is built from a combination of the LXDE and Razor-qt projects. Linux distribution offers a version with LXQt as the default desktop, which includes Artix Linux LXQt program, Lubuntu, Manjaro LXQt program, LXQt spin for Fedora Linux, and SparkyLinux LXQt Full Edition, while other similar distributions – Debian and openSUSE – offer you another desktop during installation.

DOI: 10.1201/9781003308676-9

History

LXDE developer Hong Jen Yee tried Qt in early 2013 and released the first version of Qt-based PCMan File Manager on March 26, 2013. He clarified, however, that this means no departure from GTK to LXDE, saying that "GTK and Qt versions will stay together." He later installed the Xrandr front-end LXDE on Qt.

On July 3, 2013, Hong Jen Yee announced the Qt port of the full LXDE suite, and on July 21, 2013, Razor-qt and LXDE announced their decision to merge the two projects. This integration meant that GTK and Qt versions lasted only a short time, but eventually the GTK version development was canceled, and all efforts were focused on the Qt port. The combination of LXDE-Qt and Razor-qt was renamed LXQt, and the first release, version 0.7.0, was made available on May 7, 2014.

With the release of version 0.13 on May 21, 2018, the LXQt project was officially separated from LXDE by moving to a separate GitHub organization. It uses Openbox as its default window manager.

In order to differentiate between LXDE and LXQt, we must first talk about a toolkit. The toolkit provides a way to draw the app's visual interface consistently. In addition to the toolkit, engineers should create and configure toolbar buttons and drop-down menus for each application. On Linux, there are two main tools: GTK + and Qt.

LXDE uses GTK + 2, which is a very old code. GTK + 3 has been around since 2011. LXDE Maintenance Hong Jen Yee had problems with some changes to the GTK + 3, so he released a GT-based hole in 2013. Shortly after that, a Qt LXDE version with a separate desktop interface called Razor-qt merged to form LXQt.

What Is LXQt?

Desktop location is what you see on screen. The panel is a way to organize applications in windows and allows you to move them around.

Windows and macOS all have desktop space. On Linux, there are multiple desktop areas. You can change desktop interactions using the same program, background library, and Linux kernel. Most Linux-based applications prefer a desktop environment that will be used automatically. Some allow you to select your favorite desktop location, while others are outside the desktop area. There is a human personality called Lubuntu that provides a desktop environment for LXQt. There is also the LXQt version of Fedora. If you are using another Linux-based operating system, you must install LXQt.

How LXQt Works

LXQt has a standard format for Windows users. The launcher application is located on the left. The system tray is at the far right. An open window appears in sequence between the launcher and the system tray. The launcher application contains the necessary components to start the program. Categories containing installed applications appear at the top, then you have system options, user session controls, and a search bar.

The interface is highly customizable and so you can change the desktop, app, and the icon theme. You can move the control panel to any side of the screen and arrange things the way you like. LXQt views all parts of the table as a widget. The default widget provides the ability to save favorite apps to the control panel, switch between multiple workspaces, and hide windows to display the desktop. It also comes with a few additional widgets, like a CPU monitor and a color selector.

One thing that makes LXQt attractive is the lack of dependence (the need to install background resources to launch the program) and the use of flexible components. For example, LXQt uses the Openbox window manager. You can use themes associated with Openbox to change the "look" of a window title. You can also adjust the button layout in the title bar. LXQt acts as a virtual desktop. It controls the desktop, does not attempt to control the entire startup process, and shuts down.

Who Should Use LXQt

There are a few important reasons for using LXQt:

- LXQt is very simple. If you want a simple desktop interaction that uses a few system resources, use LXQt.

- LXQt is based on Qt. Not many Qt-based desktop environments are possible compared to GTK +. If you like the Qt app but are not a fan of KDE Plasma Desktop, LXQt is one of your few options.

- LXQt is modular. If you do not want a desktop that tries to do everything, LXQt is the right choice.

- LXQt does not receive as much attention as other desktop locations. That doesn't mean it's not good. But if you are looking for alternatives, these are the simplest Linux distribution sites you can find.

SOFTWARE COMPONENTS

LXQt consists of many software components depending on Qt and KDE Frameworks 5.

Qterminal

QTerminal is a lightweight Qt terminal simulator based on QTermWidget. It is the ultimate open-source emulator, specially designed for Linux distribution, built with advanced features including, split end, multiple tab, custom shortcut, and unique color scheme.

The only lxqt-build-build tool bonds represent the architectural dependence. All major Linux and BSD platforms offer the official binary packages. Just use package manager to search for the qterminal character unit.

Falkon

Falkon is a KDE web browser that uses the QtWebEngine rendering engine, known as QupZilla. It aims to be a lightweight browser available in all major forums. This project was initially started for educational purposes only. But from its inception, Falkon has grown into a rich browser. Falkon has all the standard functions you expect in a web browser. It includes bookmarks, history (both and sidebar), and tabs. It automatically enables you to block ads with the built-in AdBlock plugin.

History

The first version of QupZilla was released in December 2010 and was written in Python with PyQt4 binding. After a few versions, QupZilla is completely rewritten in C ++ with Qt Framework. The first public release was 1.0.0-b4. Up to version 2.0, QupZilla used QtWebKit. QtWebKit has now been withdrawn, and newer versions use QtWebEngine.

sddm

Simple Desktop Display Manager (SDDM) is the display manager (image entry program and session manager) for X11 and Wayland window systems. SDDM was written from scratch in C ++ 11 and supported themes with QML. SDDM is a free and open-source software subject to GNU General Public License version 2 or later.

lximage-qt

LXImage-Qt is a Qt port for LXImage, a simple and fast image viewer.

Features

- Zoom in, rotate, rotate and resize images

- Slide show

- Thumbnail bar (left, top, or bottom); different icon sizes

- Exif data bar

- Rename the linear image

- Custom shortcuts

- Picture annotations (arrow, rectangle, circle, numbers)

- Recent files

- Upload photos (Imgur, ImgBB)

- Take screenshots

Some features can be obtained when used. LXImage-Qt is maintained by the LXQt project but can be used independently in this desktop environment.

lxmenu data

The LXMenu Data Package provides the files needed to create desktop menus that are compatible with LXDE Freedesktop.org. This package is known for its build and performance using the LFS-11.0 platform.

lxqt-about

It is a chat window that provides information about LXQt and its operating system. The library is provided by all other major Linux components such as Arch Linux, Debian, and openSUSE. Just use the package manager to search for the lxqt-about character unit.

lxqt-admin

This repository provides two GUI tools to configure LXQt operating system settings.

Usage

Like the same tools provided by lxqt-config, lxqt-admin tools can be launched from the Configuration Center and in the main menu – Favorites – LXQt settings. Real use should be self-explanatory. Using settings GUI verification agent for active polkit is introduced to get root password.

lxqt-archive

Simple and easy Qt file storage. The main I/O functions are transferred from Engrampa (Gnome File Roller fork). This is just the result (archive) of archiving programs such as tar and zip.

The supported file types are as follows:

1. 7-Zip Compressed File (.7z)

2. WinAce Compressed File (.ace)

3. Compressed ALZip File (.alz)

4. Archive with a small AIX (.ar) index

5. Compressed ARJ archive (.arj)

6. Cabinet File (.cab)

7. UNIX CPIO Archive (.cpio)

8. Debian Linux Package (.deb) [Read Mode Only]

9. ISO-9660 CD Disc Image (.iso) [Read Mode Only]

10. Java archive (.jar)

11. Java Enterprise (.ear) Archive

12. Java (Archive) web archive

13. LHA archive (.lzh, .lha)

14. Depressed WinRAR archive (.rar)

15. RAR Archived Comic Book (.cbr)

16. RPM Linux Package (.rpm) [Read Mode Only]

17. Saved tape file: * uncompressed (.tar) or compressed with: * gzip (.tar .gz, .tgz) * bzip (.tar.bz, .tbz) * bzip2 (.tar.bz2, .tbz2) * press (.tar.Z,

.taz) * lrzip (.tar.lrz, .tlrz) * lzip (.tar.lz, .tlz) * lzop (.tar.lzo, .tzo) * 7zip (.tar.7z) * xz (.tar.xz)

18. Object Archives (.bin, .sit)

19. ZIP archive (.zip)

20. ZIP Archive Archive (.cbz)

21. Compressed ZOO Archive File (.zoo)

22. Individual files are compressed with gzip, bzip, bzip2, compress, lrzip, lzip, lzop, rzip, xz.

lxqt-common

This collection includes a number of support files used by various LXQt components. Among these are image files, themes, desktop installation files according to the XDG Desktop Menu Definition, template configuration files in various components such as PCManFM-Qt or Openbox window manager, and script startlxqt used to start LXQt times. The LXQt logo was designed by @ Caig and licensed CC-BY-SA 3.0. The LXQt theme "Plasma" is based on the Next KDE Plasma theme by the KDE Visual team.

The Openbox window manager automatically stores user settings in the $ XDG_CONFIG_HOME / openbox / rc.xml file, usually / home / <user> /.config/openbox/rc.xml. When Openbox is used as the LXQt window manager, the $ XDG_CONFIG_HOME / openbox / lxqt-rc.xml file is used instead. This allows you to keep LXQt-specific settings simultaneously using different settings when Openbox is e. g. is used in "only" sessions of a private window manager outside of any desktop area. LXDE uses the same way the custom configuration file $ XDG_CONFIG_HOME / openbox / lxde-rc.xml.

In order to maintain a consistent return, those configuration files are handled by LXQt as follows:

1. Lxqt-common sends template file $ XDG_CONFIG_DIRS / openbox / lxqt-rc.xml, usually /etc/xdg/openbox/lxqt-rc.xml.

2. At the beginning of each LXQt session, startlxqt checks that any rc.xml, lxde-rc.xml, or lxqt-rc.xml files are available at $ XDG_CONFIG_HOME / openbox /.

3. Either way this will result in the file $ XDG_CONFIG_HOME / openbox / lxqt-rc.xml available for use from then on during LXQt periods.

lxqt-config

This repository provides several tools for configuring both the LXQt and the underlying operating system. On the other hand, it has several GUI tools to customize topics such as standard appearance, interface devices, or screen resolutions. On the other hand, the GUI "Configuration Center" summarizes all those configuration tools and counterparts of other LXQt components or third-party applications.

GUI CONFIGURATION TOOLS

LXQt Appearance Configuration

LXQt look includes titles such as the LXQt icon and theme or fonts. Binary lxqt-config-look.

Brightness

Brightness settings for output devices. Technically, the colors are adjusted to mimic different brightness if the LXQt operating system does not allow you to adjust the light itself.

File Associations

Provides MIME types for applications used to handle them. Binary lxqt-config-file-associations.

Keyboard and Mouse

Configuring the hardware of the devices. Settings such as repeat delays and keyboard intervals or device speed acceleration.

Locale

Locale used within the LXQt session. This GUI sets the natural variables known as LANG or LC_ *. Settings apply to the entire session, i.e., to programs running within LXQt sessions but not to LXQt sessions, too. Binary lxqt-config-locale.

Monitor Settings

Adjusts screen resolutions, screen layouts, and other preferences.

lxqt-globalkeys

This room provides tools for setting up global keyboard shortcuts for LXQt sessions, which run for the entire LXQt session and are not limited to separate applications. The main components are two lxqt-globalkeysd binary and lxqt-config-globalkeyshortcuts. Lxqt-globalkeysd works similarly to a daemon called the LXQt Module and does real work. The GUI lxqt-config-globalkeyshortcuts is used to customize the shortcut settings.

Binary Packages

Official binary packages are offered by all major Linux and BSD shares. Just use package manager to search for lxqt-globalkeys.

Usage

Windows administrators can provide shortcuts and also their range may exceed one of the lxqt-global keys in LXQt sessions. It resulted in a warning message "Global shortcut C + A + d + cannot be registered," see https://github.com/lxqt/lxqt/issues/1032. As shown in this version, lxqt-notificationd will display a warning in the event of such a conflict. Users can decide if they want the said shortcut handled by the appropriate window manager lxqt-globalkeys keys.

lxqt-globalkeys

Daemon-like lxqt-globalkeysd can be modified from the "Basic Settings" section in the LXQt Session Settings session setting for lxqt-session. Configuration dialog "Global Activity Manager" (binary lxqt-config-globalkey shortcuts) used to customize shortcuts can be opened in the main menu panel – Preferences – LXQt Settings – Shortcut Keys and provided by the lxqt-config Setup Center.

lxqt-notification

Lxqt-notificationd is the use of the LXQt daemon according to the Desktop Notifications Specification. This specification explains how to display notifications in pop-up windows on desktops. Notices like these are frequently used in chat or email clients to notify incoming messages, media players to signal the start of another track but also on the desktop itself to indicate volume changes or the like. Lxqt-notificationd combines dual lxqt-notificationd and lxqt-config-notificationd. Lxqt-notificationd works in a similar way to a daemon called the LXQt Module and performs a real function. The "Desktop Notifications" GUI, binary lxqt-config-notificationd, is used to customize notifications. Lxqt-notificationd uses version 1.2 of this expression.

lxqt-qtplugin

This repository provides library libqtlxqt to integrate Qt and LXQt. With this plugin, all Qt-based programs can use LXQt settings, as a thumbnail theme. Official Binary Packages are offered by all major Linux components such as Arch Linux, Debian (like Debian stretch only), Fedora, and -openSUSE. Just use your package manager to search for the lxqt-qtplugin character unit.

lxqt-panel

The LXQt panel is the default bar at the bottom of your screen. The panel contains an application menu, desktop switch, instant launch bar, task bar, system tray containing applets, and clock. The panel is highly customizable with several settings and plugins detailed in the customization section and also contains how to add or remove apples.

lxqt-runner

Lxqt-runner provides GUI from desktop and allows application launch or program shutdown. The calculation function is also used. Technically, it contains a binary lxqt-runner. Binary is used similarly to a daemon called the LXQt Module and delivers the GUI on hitting the default keyboard shortcut.

lxqt-session

Repository lxqt-session provides tools for handling LXQt sessions. First there is the lxqt-session session manager. Binary was introduced at the beginning of the LXQt session and is responsible for introducing and monitoring all the other components that make up the sessions. The GUI "LXQt Session Settings" (lxqt-config-session binary) is used to adjust various settings that affect the session, e.g., any windows manager can use any applications that should start automatically. Binary lxqt-leve is about interrupting or interrupting times. It comes with a few options that reflect the action we are starting, e.g., --leave, --hibernate, --reable, or --restart.

INSTALLING LXQT DESKTOP ON UBUNTU

Run the following command to install LXQt desktop given below.

```
$ sudo apt install lxqt sddm
```

After installation, reboot your Ubuntu system. Select desktop session as LXQt Desktop. You should see the welcome screen when you select for it.

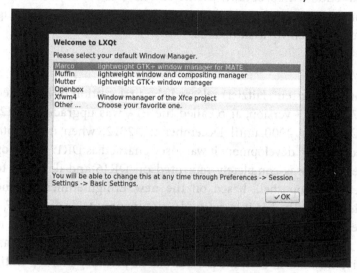

Welcome Screen of LXQt.

The first window of the LXQt appears as

LXQt Desktop.

ENLIGHTENMENT DESKTOP ENVIRONMENT

Light, also known as E, is a compact window manager for the X Window System. From version 20, Enlightenment is also the creator of Wayland. Light developers have dubbed it the "real eye candy window manager." Enlightenment includes image shell rendering functions and can be used in conjunction with programs written for GNOME or KDE. Used in conjunction with the Enlightenment Foundation Libraries (EFL), Lighting can refer to the entire desktop area.

History

Rasterman (Carsten Haitzler) released the first edition of Enlightenment in 1997. The 0.17 version, also called the E17, was upgraded for 12 years from December 2000 until December 21, 2012, when it was officially released. During development it was also renamed as DR17 (Development Release 17). It was completely reprinted on DR16 and designed to be a complete desktop shell, based on the new Enlightenment Foundation Libraries (EFL). The E16 is in active development outside of the E17, reaching a record 1.0 in 2009 (1.0.23 from 2021). Bodhi Linux was built next to the Enlightenment 17 desktop, but put it on a fork to build a Moksha desktop. Elive Linux also used the E17 fork as its primary desktop until 2019, when the 3.7 series was built. The current version is E25.

Version

Here is the list of versions of the enlightenment desktop environment with its features as given below,

E16

E16 provides features that allow users to create a grid of workspaces known as "virtual desktops." It allows switching between them, which is achieved by hurling the mouse cursor. You can have up to 8 × 8 desktops in a single grid, and up to 32 grids are possible, making 2,048 possible workspaces. You can enable a sort of "map" of the desktops, called the pager if they get lost. It also provides the ability to put windows in groups "iconification," which is similar to minimizing but the windows are stored in "iconboxes" that can be moved, the ability to change the type of or remove altogether the borders and title bars, advanced keybinding settings to allow the use of Enlightenment solely with a keyboard a compositor with effects such as fading and transparency.

E17

The E17 has many key features such as the following:

1. It has a full theme, in both a menu-based and a command line that changes the visual theme

2. Built-in file manager

3. Thumbnails on desktop

4. The visible desktop grid feature

5. One or more shelves for handling gadget placement and on-screen appearance

6. Animated, interactive desktop background, menu items, iBar items, and desktop widgets are all possible

7. Window blurring, thumbnails, enlargement, and attachment settings

8. Customized key binding

9. Foreign trade support

10. Level: supports all required standards (NetWM, ICCCM, XDG, and more)

E21

1. Wayland support has dramatically improved

2. New gadget infrastructure

3. Wizard development

4. In the background of the video

E22

1. Wayland's support has dramatically improved.

2. Development of new gadget infrastructure

3. Added sudo / ssh askpass utility GUI

4. Meson building system

5. Tile policy development

6. Integrated volume controls per window

E23

1. New decorated screen option

2. Meson now builds a building program.

3. Music Control supports rage MPRIS dbus protocol

4. Add Bluez5 support with a new module and completely redesigned gadget

5. Add dpms option to close or unlock it

6. Alt-tab window switch allows windows to move while alt configuration

7. Multiple bug fixes include alerts, etc.

8. Significant improvements in Wayland support

INSTALLING ENLIGHTENMENT ON UBUNTU

Run the following command to install LXQt desktop given below.

```
$ sudo apt-get install enlightenment
```

After installation, reboot your Ubuntu system. Select desktop session as LXQt Desktop. The first window of the LXQt will appear as

Enlightenment Desktop.

PANTHEON DESKTOP ENVIRONMENT

Pantheon contains everything you see on screen. It shows the background of your desktop that switches between open windows, the home screen at the top of the screen, and the dock at the bottom. Pantheon is one of the many available Linux desktop environments. This situation is different from Windows and macOS, each giving you one interface to use. That's why most of us have no idea that the desktop environment is separate from the kernel, bootloader, or any other part of those two apps. If you can't change the parts, the difference doesn't matter.

History

Unlike many other desktop environments, Pantheon is very close to the Linux operating system (more commonly known as "distribution" or "distro"). The engineers behind Elementary OS are the same people behind the Pantheon. The Pantheon started as another way people could put it on Ubuntu. Ubuntu is widely regarded as the most popular Linux version of personal computers if that name is unfamiliar. At that time, it used a desktop space known as Unity. That link had a panel at the top and a separate icon dock at the bottom left of the screen. The design encouraged people to find and unlock the software by searching.

The Pantheon was no other way to Unity than GNOME, one of the oldest and most established desktop centers for free desktop and open source. Elementary OS founder Daniel Fore and others knew they could not change GNOME to their favorite, so they used multiple building blocks to build something of their own. Their creation, Pantheon, is written with GTK + and Vala. Unlike other developed methods like GNOME and the KDE Plasma desktop, which you can install on almost any Linux version, the Pantheon is most visible on Elementary OS only. That being said, you do not need to use Elementary OS to install the Pantheon. As a free project, some can use and redistribute the code as they wish.

The Pantheon desktop environment is built on the GNOME software base, GTK, GDK, Cairo, GLib (GObject and GIO), GVfs, and Tracker. The desktop allows multiple workplaces to schedule user activity. Pantheon apps designed and developed by foundations include:

- **Pantheon Greeter:** LightDM based session manager

- **Gala:** Window manager

- **Wingpanel:** Top panel, similar to working with the top GNOME Shell panel

- **Slingshot:** An app launcher available on WingPanel
- **Wood:** Dock (where Docky is based)
- **Switchboard:** Settings application (or control panel)
- **Pantheon Mail:** Email client written in Close and supported by WebKitGTK
- **Calendar:** Desktop calendar
- **Music:** Audio Player
- **Code:** A text-based editor comparable to gedit or leafpad.
- **Terminal:** Terminal emulator
- **Files (formerly called Marlin):** File manager
- **Installer:** Installer built in conjunction with System76

Bryan Lunduke of Network World wrote that the Pantheon desktop environment, the first OS platform, was among the best of 2016. Pantheon is also being released as an ongoing desktop of choice of GeckoLinux.

Development

- Jupiter
- 0.2 Luna
- 0.3 Freya
- 0.4 Loki
- 5.0 Juno
- 5.1 Hera
- Odin
- 6.1 Jólnir

Features
The New Interface
Starting with the user interface, we discover a new Pantheon desktop. Pantheon is a home desktop explicitly designed for the first OS, which

has gained a lot of flow. This clean and modern desktop may be one of the main reasons for the initial success.

Dark Mode

Pantheon finally gets black mode. Everything looks clean and advanced on the first OS while all the operating system applications and applications installed to follow this black mode maintenance process is made so as not to force third-party applications to use this black mode which could lead to a bad UI. Black mode is just recommended. You can also set a dark mode to enter at sunset or from time to time.

Themes and Looks

The first OS 6 also brings advanced customization with ten different colors to choose from. This is very similar to changing themes. But this will be greatly polished. You can also choose the primary color behind the desktop. Before the customization area, Odin in the first os was close to zero.

Performance

While playing with the original OS 6, we noticed a slight increase in operating times. Most automated system apps also have Flatpacks and Flatpacks with the original head up. They have all their dependencies attached to them. They also have to be read and uploaded to memory when you click the app icon.

Usability and Stability

The first OS has been stable and reliable. It is based on the solid foundation of Ubuntu and is a great program for homes, students, offices, software developers, and businesses. The foundation was created with a high level of technology and is visible everywhere. The first OS 6 Odin will be supported for the next 4+ years. It will continue to receive regular bug fix updates and security updates.

CHAPTER SUMMARY

In this chapter, we have covered three DE introductions and its features, history, core projects, applications, and version history. We have provided a separate section where you get history of their versions.

Appraisal

UBUNTU AND OTHER LINUX distro systems took the open-source and the IT world by storm in the early years. They support various desktop environments such as GNOME, KDE, MATE, and more. The operating system could have grown into a fully featured desktop and server offering that has won users' hearts globally. Besides the solid technical platform and impressive dedication to quality, desktop environment (DE), e.g., MATE, GNOME, LXDE, KDE, enjoys success because of its vast community of enthusiastic users who help support, document, and test every point of Linux landscape. The term used in this chapter is the desktop environment.

So what is DE? A desktop environment is the desktop metaphor of a bundle of programs running on top of an operating system that shares a standard graphical user interface (GUI). It is described as a graphical shell. The desktop environment mainly was on personal computers until mobile computing. Desktop GUIs help the user quickly access and edit files, while they usually don't provide access to all of the features in the underlying operating system. Besides, the traditional command-line interface (CLI) is still used when complete control over the operating system is required.

It consists of various icons, windows, toolbars, folders, wallpapers, and desktop widgets. A GUI provides drag and drop functionality and other features that complete the desktop metaphor. A desktop environment aims to be a way for the user to interact with the system using concepts similar to those used to interact with the rest of the world, such as buttons and windows. This is the piece users are interacting with. There are many desktop areas (GNOME, Cinnamon, Mate, Pantheon, Enlightenment, KDE, Xfce, etc.). Each desktop includes built-in applications (such as file managers, configuration tools, web browsers, and games).

You have the official guide to these unique desktop environments in your hands. Each of us working on this book has shown a high level of technical competence and shares this knowledge with you all. We gathered together to create a book that offers a solid understanding of the essential points of Ubuntu and explains the fundamentals along with the other LXDE, Budgie, Cinnamon, or other details.

This book varies in the coverage of various topics. This is intentional. Other books do not cover some topics, and those topics deserve more coverage here. There are some topics that power users master. Other topics are things the power users should know about. They can understand some history, some other options, or have what they need to listen to further discussions with different technical views without being completely confused.

This book is planned for intermediate and advanced users or those who want to become middle and advanced users. Our goal is to give you the right direction, to help you enter the higher stages by telling you to use as many tools and ideas as possible. We give you some thoughts and methods to consider so that you can seek out more. Although the content of this book is for intermediate to advanced users, new users who pay attention will benefit from each chapter as all chapters are related. The central pointer is that more detailed or related information is provided at the end of each chapter.

This book helps you to learn these skills and tells you how to learn more about your system, Linux, with the software including Ubuntu Distros with the desktop environment. Most importantly, it enables you to overcome your fear of the system by telling you more about it and how it works. You can also install other Linux distros like Fedora, OpenSuse, Manjora, etc.

This book is not a pure reference book but properly guides you with step-by-step procedures for performing tasks. This book is organized by topics and includes many useful commands.

Chapter 1 discusses the basic understanding of the desktop environment, and its features also have some specific terms like GUI, CLI, TUI. The chapter describes the vast resources available to support this book. You will also get a brief knowledge of DE history, features, and some pros and cons.

Chapter 2 gives a quick review of the KDE Plasma introduction installation, describes valuable commands such as apt-get snapd, and gives some brief knowledge of the user interface, core projects, and version history environment system.

Chapter 3 provides a quick review of the GNOME introduction, installation, GNOME-based distros, pros, and cons, and also gives some brief knowledge of the GNOME user interface.

Chapter 4 discusses XFCE desktop environment, its versions, history, main components of XFCE, installation, advantages, and disadvantages.

Chapter 5 discusses other MATE desktop environments, version history, main components of MATE, installation, advantages and disadvantages, and another operating system for MATE.

Chapter 6 provides you with knowledge of the other Budgie desktop environment, its version history, main components of Budgie, installation, advantages, and disadvantages.

Chapter 7 discusses Cinnamon, its version history, main components of Cinnamon, libraries, core components, installation, advantages, and disadvantages.

Chapter 8 discusses LXDE, its version history, main components of Cinnamon, libraries, core components of the software, reasons to use LXDE, installation, also with Lubuntu distribution.

Chapter 9 discusses other DEs such as Pantheon, Enlightenment, and LXQt, version history, main components, and features.

In other words, this book provides a Linux desktop environment on any Linux-based distribution system.

Bibliography

Budgie (Desktop Environment). (2013, December 7). Wikipedia. https://en.wikipedia.org/wiki/Budgie_(desktop_environment). Last edited on October 5, 2022.

Cinnamon (Desktop Environment). (2011, January 1). Wikipedia. https://en.wikipedia.org/wiki/Cinnamon_(desktop_environment). Last edited on September 21, 2022.

Cinnamon. (2012, November 12). Wikipedia. https://en.wikipedia.org/wiki/Cinnamon. Last edited on October 14, 2022.

Cinnamon. (n.d.). Linux Mint Developer Guide Documentation. Retrieved July 11, 2022, from https://linuxmint-developer-guide.readthedocs.io/en/latest/cinnamon.html

Comparing Graphical User Interface (GUI) and Command Line Interface (CLI). (n.d.). Engineering Education (EngEd) Program | Section. Retrieved July 11, 2022, from https://www.section.io/engineering-education/comparing-graphical-user-interface-gui-and-command-line-interface-cli/

Desktop Environment. (2012, February 4). Wikipedia. https://en.wikipedia.org/wiki/Desktop_environment. Last edited on October 11, 2022.

Development/Tutorials/Using KParts. (n.d.). KDE TechBase. Retrieved July 11, 2022, from https://techbase.kde.org/Development/Tutorials/Using_KParts#:~:text=KPart%20technology%20is%20used%20in%20kde%20to%20reuse,just%20use%20a%20katepart%20or%20a%20konsolepart%20instead

Emms, S. (2021, July 11). *LXQt - The Lightweight Qt Desktop Environment*. LinuxLinks. https://www.linuxlinks.com/lxqt-lightweight-qt-desktop-environment/

GNOME. (1999, March 3). Wikipedia. https://en.wikipedia.org/wiki/GNOME. Last edited on October 11, 2022.

GNOME Foundation. (1999, March 3). GNOME 1.0 Released, Press Release. https://foundation.gnome.org/1999/03/03/gnome-1-0-released/

Graphical User Interface. (2022, May 1). Wikipedia. https://en.wikipedia.org/wiki/Graphical_user_interface#:~:text=The%20graphical%20user%20interface%20(GUI,command%20labels%20or%20text%20navigation. Last edited on October 11, 2022.

Is Ubuntu based on GNU/Linux? (n.d.). Quora. Retrieved July 11, 2022, from https://www.quora.com/Is-Ubuntu-based-on-GNU-Linux

K Desktop Environment 1. (1998, July 12). Wikipedia. https://en.wikipedia.org/wiki/K_Desktop_Environment_1. Last edited on February 27, 2022.

K Desktop Environment 2. (2000, October 23). Wikipedia. https://en.wikipedia.org/wiki/K_Desktop_Environment_2. Last edited on February 27, 2022.

K Desktop Environment 3. (2002, April 3). Wikipedia. https://en.wikipedia.org/wiki/K_Desktop_Environment_3. Last edited on May 27, 2022.

KDE Applications. (n.d.). Retrieved July 11, 2022, from https://apps.kde.org/

KDE Plasma 5. (2014, July 15). Wikipedia. https://en.wikipedia.org/wiki/KDE_Plasma_5. Last edited on October 29, 2022.

KDE Projects. (2010, November 6). Wikipedia. https://en.wikipedia.org/wiki/KDE_Projects. Last edited on September 30, 2022.

KDE Software Compilation 4. (2008, January 11). Wikipedia. https://en.wikipedia.org/wiki/KDE_Software_Compilation_4. Last edited on August 22, 2022.

KDE UserBase Wiki. (n.d.). Retrieved July 11, 2022, from https://userbase.kde.org/Welcome_to_KDE_UserBase. Last edited on November 26, 2020.

KHTML. (2021, May 2). Wikipedia. https://en.wikipedia.org/wiki/KHTML. Last edited on September 6, 2022.

Kili, A. (2016, August 31). *10 Best and Most Popular Linux Desktop Environments of All Time.* https://www.tecmint.com/best-linux-desktop-environments/

KIO. (2022, June 27). Wikipedia. https://en.wikipedia.org/wiki/KIO. Last edited on April 1, 2022.

KJS (software). (2000, January 1). Wikipedia. https://en.wikipedia.org/wiki/KJS_(software). Last edited on February 19, 2022.

List of Graphical User Interface Elements. (2006, September 19). Wikipedia. https://en.wikipedia.org/wiki/List_of_graphical_user_interface_elements. Last edited on April 1, 2022.

LXDE. (2006, January 1). Wikipedia. https://en.wikipedia.org/wiki/LXDE. Last edited on August 13, 2022.

LXDE. (n.d.). ArchWiki. Retrieved July 11, 2022, from https://wiki.archlinux.org/title/LXDE

LXQt - The Lightweight Qt Desktop Environment. (n.d.). Retrieved July 11, 2022, from https://lxqt-project.org/

LXQt. (2013, January 1). Wikipedia. https://en.wikipedia.org/wiki/LXQt. Last edited on August 18, 2022.

LXQt. (n.d.). The Lightweight Qt Desktop Environment. Retrieved July 11, 2022, from https://lxqt-project.org/

Module Solid. (n.d.). Retrieved July 11, 2022, from https://api.kde.org/legacy/pykde-4.1-api/solid/index.html

Plasma. (2022, June 10). KDE Community. https://kde.org/plasma-desktop/

Software:Cinnamon (Desktop Environment). (2011, January 1). HandWiki. https://handwiki.org/wiki/Software:Cinnamon_(desktop_environment)

Software:ThreadWeaver. (2021, August 13). HandWiki. https://handwiki.org/wiki/Software:ThreadWeaver

Sonnet. (n.d.). KDE Community Wiki. Retrieved July 11, 2022, from https://community.kde.org/Sonnet

Team, T. M. (2013, October 31). *MATE Desktop Environment.* https://mate-desk-top.org/

Text-based User Interface. (2014, September 1). Wikipedia. https://en.wikipedia.org/wiki/Text-based_user_interface. Last edited on October 12, 2022.

Ubuntu Budgie. (2016, April 25). Wikipedia. https://en.wikipedia.org/wiki/Ubuntu_Budgie. Last edited on July 28 2022.

Ubuntu Budgie. (n.d.). Retrieved July 11, 2022, from https://ubuntubudgie.org/

Ubuntu GNOME. (2012, October 18). Wikipedia. https://en.wikipedia.org/wiki/Ubuntu_GNOME. Last edited on July 21, 2022.

What is a GUI (Graphical User Interface)? (n.d.). Retrieved July 11, 2022, from https://www.computerhope.com/jargon/g/gui.htm

What is Free Software?- GNU Project. (n.d.). Free Software Foundation. Retrieved July 11, 2022, from https://www.gnu.org/philosophy/free-sw.en.html

What is Linux?. (n.d.). Linux.Com. Retrieved July 11, 2022, from https://www.linux.com/what-is-linux/

Wikizero. (2008, January 11). Phonon (Software). https://wikizero.com/index.php/en/Phonon_(KDE)

X Power Tools. (n.d.). O'Reilly Online Learning. Retrieved July 11, 2022, from https://www.oreilly.com/library/view/x-power-tools/9780596101954/ch01.html

X Power Tools. (n.d.). O'Reilly Online Learning. Retrieved July 11, 2022, from https://www.oreilly.com/library/view/x-power-tools/9780596101954/ch01.html

X Window System. (1984, June 1). Wikipedia. https://en.wikipedia.org/wiki/X_Window_System. Last edited on September 21, 2022.

Xerox Alto. (1973, March 1). Wikipedia. https://en.wikipedia.org/wiki/Xerox_Alto. Last edited on September 28, 2022.

Xfce Desktop Environment. (2020, December 1). https://xfce.org/

Xfce. (1996, January 1). Wikipedia. https://en.wikipedia.org/wiki/Xfce. Last edited on October 11, 2022.

XMLGUI. (2017, April 1). Wikipedia. https://en.wikipedia.org/wiki/XMLGUI. Last edited on December 23, 2018.

Index

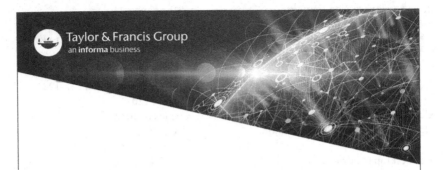

Printed in the United States
by Baker & Taylor Publisher Services